Introduction to Linear Algebra

Introduction to Linear Algebra

Edited by
Lilian Mandelbaum

Willford Press
www.willfordpress.com

Published by Willford Press,
118-35 Queens Blvd., Suite 400,
Forest Hills, NY 11375, USA

ISBN: 978-1-68285-491-4

Cataloging-in-Publication Data

Introduction to linear algebra / edited by Lilian Mandelbaum.
p. cm.
Includes bibliographical references and index.
ISBN 978-1-68285-491-4
1. Algebras, Linear. 2. Mathematics. I. Mandelbaum, Lilian.
QA184.2 .I58 2018
512.5--dc23

For information on all Willford Press publications
visit our website at www.willfordpress.com

Willford Press

Contents

Preface

Linear algebra refers to the study of lines, sub-spaces, planes and understanding the characteristics of vector spaces. It includes learning and analysing linear mapping. Linear algebra is also important in understanding differential equations, abstract algebra, functional analysis, etc. It has applications in fourier series expansion, linear systems, quantum mechanics, etc. This book presents the complex subject of linear algebra in the most comprehensible and easy to understand language. For someone with an interest and eye for detail, this text covers the most significant topics in the field of linear algebra. The textbook is appropriate for those seeking detailed information in this area.

To facilitate a deeper understanding of the contents of this book a short introduction of every chapter is written below:

Chapter 1- The branch of mathematics which studies vector spaces is known as linear algebra. Natural sciences, computer science, analytic geometry and computer animation are some of the subjects where techniques from linear algebra can be applied. The chapter on linear algebra offers an insightful focus, keeping in mind the complex subject matter.

Chapter 2- Inner product space is a vector space along with a function called inner product. An inner product is also a normed vector space as it produces an associated norm. The topics discussed in the chapter are of great importance to broaden the existing knowledge on vector space.

Chapter 3- A matrix can be defined as an array of numbers, symbols, etc. that are arranged in rows and columns. Matrices can be applied in a number of fields like classical mechanics, quantum mechanics, electromagnetism and quantum electromagnetism. The aspects elucidated in this chapter are of vital importance, and provide a better understanding of matrices.

Chapter 4- A linear transformation between two vector spaces $V \rightarrow W$ can be such that the operations of addition and scalar multiplication are preserved. The expected value of a random variable, the identity map on any module are examples of linear values. Linear algebra is best understood in confluence with the major topics listed in the following chapter.

Finally, I would like to thank the entire team involved in the inception of this book for their valuable time and contribution. This book would not have been possible without their efforts. I would also like to thank my friends and family for their constant support.

Editor

1

An Overview of Linear Algebra

The branch of mathematics which studies vector spaces is known as linear algebra. Natural sciences, computer science, analytic geometry and computer animation are some of the subjects where techniques from linear algebra can be applied. The chapter on linear algebra offers an insightful focus, keeping in mind the complex subject matter.

Linear Algebra

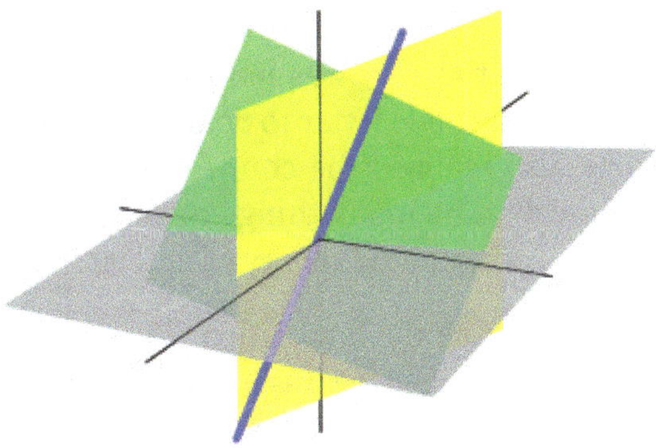

The three-dimensional Euclidean space R^3 is a vector space, and lines and planes passing through the origin are vector subspaces in R^3.

Linear algebra is the branch of mathematics concerning vector spaces and linear mappings between such spaces. It includes the study of lines, planes, and subspaces, but is also concerned with properties common to all vector spaces.

The set of points with coordinates that satisfy a linear equation forms a hyperplane in an n-dimensional space. The conditions under which a set of n hyperplanes intersect in a single point is an important focus of study in linear algebra. Such an investigation is initially motivated by a system of linear equations containing several unknowns. Such equations are naturally represented using the formalism of matrices and vectors.

Linear algebra is central to both pure and applied mathematics. For instance, abstract algebra arises by relaxing the axioms of a vector space, leading to a number of generalizations. Functional analysis studies the infinite-dimensional version of the theory of vector spaces. Combined with calculus, linear algebra facilitates the solution of linear systems of differential equations.

Techniques from linear algebra are also used in analytic geometry, engineering, physics, natural sciences, computer science, computer animation, advanced facial recognition algorithms and the social sciences (particularly in economics). Because linear algebra is such a well-developed theory, nonlinear mathematical models are sometimes approximated by linear models.

History

The study of linear algebra first emerged from the study of determinants, which were used to solve systems of linear equations. Determinants were used by Leibniz in 1693, and subsequently, Gabriel Cramer devised Cramer's Rule for solving linear systems in 1750. Later, Gauss further developed the theory of solving linear systems by using Gaussian elimination, which was initially listed as an advancement in geodesy.

The study of matrix algebra first emerged in England in the mid-1800s. In 1844 Hermann Grassmann published his "Theory of Extension" which included foundational new topics of what is today called linear algebra. In 1848, James Joseph Sylvester introduced the term matrix, which is Latin for "womb". While studying compositions of linear transformations, Arthur Cayley was led to define matrix multiplication and inverses. Crucially, Cayley used a single letter to denote a matrix, thus treating a matrix as an aggregate object. He also realized the connection between matrices and determinants, and wrote "There would be many things to say about this theory of matrices which should, it seems to me, precede the theory of determinants".

In 1882, Hüseyin Tevfik Pasha wrote the book titled "Linear Algebra". The first modern and more precise definition of a vector space was introduced by Peano in 1888; by 1900, a theory of linear transformations of finite-dimensional vector spaces had emerged. Linear algebra took its modern form in the first half of the twentieth century, when many ideas and methods of previous centuries were generalized as abstract algebra. The use of matrices in quantum mechanics, special relativity, and statistics helped spread the subject of linear algebra beyond pure mathematics. The development of computers led to increased research in efficient algorithms for Gaussian elimination and matrix decompositions, and linear algebra became an essential tool for modelling and simulations.

Educational History

Linear algebra first appeared in American graduate textbooks in the 1940s and in undergraduate textbooks in the 1950s. Following work by the School Mathematics Study Group, U.S. high schools asked 12th grade students to do "matrix algebra, formerly reserved for college" in the 1960s. In France during the 1960s, educators attempted to teach linear algebra through finite-dimensional vector spaces in the first year of secondary school. This was met with a backlash in the 1980s that removed linear algebra from the curriculum. In 1993, the U.S.-based Linear Algebra Curriculum Study Group recommended that undergraduate linear algebra courses be given an application-based "matrix orientation" as opposed to a theoretical orientation. Reviews of the teaching of linear algebra call for stress on visualization and geometric interpretation of theoretical ideas, and to include the *jewel in the crown* of linear algebra, the singular value decomposition (SVD), as 'so many other disciplines use it'. To better suit 21st century applications, such as data mining and uncertainty analysis, linear algebra can be based upon the SVD instead of Gaussian Elimination.

Scope of Study

Vector Spaces

The main structures of linear algebra are vector spaces. A vector space over a field F (often the field of the real numbers) is a set V equipped with two binary operations satisfying the following axioms. Elements of V are called *vectors*, and elements of F are called *scalars*. The first operation, *vector addition*, takes any two vectors v and w and outputs a third vector $v + w$. The second operation, *scalar multiplication*, takes any scalar a and any vector v and outputs a new vector av. The operations of addition and multiplication in a vector space must satisfy the following axioms. In the list below, let u, v and w be arbitrary vectors in V, and a and b scalars in F.

Axiom	**Signification**
Associativity of addition	$u + (v + w) = (u + v) + w$
Commutativity of addition	$u + v = v + u$
Identity element of addition	There exists an element $0 \in V$, called the *zero vector*, such that $v + 0 = v$ for all $v \in V$.
Inverse elements of addition	For every $v \in$ V, there exists an element $-v \in V$, called the *additive inverse* of v, such that $v + (-v) = 0$
Distributivity of scalar multiplication with respect to vector addition	$a(u + v) = au + av$
Distributivity of scalar multiplication with respect to field addition	$(a + b)v = av + bv$
Compatibility of scalar multiplication with field multiplication	$a(bv) = (ab)v$
Identity element of scalar multiplication	$1v = v$, where 1 denotes the multiplicative identity in F.

The first four axioms are those of V being an abelian group under vector addition. Elements of a vector space may have various nature; for example, they can be sequences, functions, polynomials or matrices. Linear algebra is concerned with properties common to all vector spaces.

Linear Transformations

Similarly as in the theory of other algebraic structures, linear algebra studies mappings between vector spaces that preserve the vector-space structure. Given two vector spaces V and W over a field F, a linear transformation (also called linear map, linear mapping or linear operator) is a map

$$T : V \rightarrow W$$

that is compatible with addition and scalar multiplication:

$$T(u + v) = T(u) + T(v), \quad T(av) = aT(v)$$

for any vectors $u, v \in V$ and a scalar $a \in F$.

Additionally for any vectors u, $v \in V$ and scalars a, $b \in F$:

$$T(au + bv) = T(au) + T(bv) = aT(u) + bT(v)$$

When a bijective linear mapping exists between two vector spaces (that is, every vector from the second space is associated with exactly one in the first), we say that the two spaces are isomorphic. Because an isomorphism preserves linear structure, two isomorphic vector spaces are "essentially the same" from the linear algebra point of view. One essential question in linear algebra is whether a mapping is an isomorphism or not, and this question can be answered by checking if the determinant is nonzero. If a mapping is not an isomorphism, linear algebra is interested in finding its range (or image) and the set of elements that get mapped to zero, called the kernel of the mapping.

Linear transformations have geometric significance. For example, 2×2 real matrices denote standard planar mappings that preserve the origin.

Subspaces, Span, and Basis

Again, in analogue with theories of other algebraic objects, linear algebra is interested in subsets of vector spaces that are themselves vector spaces; these subsets are called linear subspaces. For example, both the range and kernel of a linear mapping are subspaces, and are thus often called the range space and the nullspace; these are important examples of subspaces. Another important way of forming a subspace is to take a linear combination of a set of vectors $v_1, v_2, \ldots, v_k$:

$$a_1 v_1 + a_2 v_2 + \cdots + a_k v_k,$$

where $a_1, a_2, \ldots, a_k$ are scalars. The set of all linear combinations of vectors $v_1, v_2, \ldots, v_k$ is called their span, which forms a subspace.

A linear combination of any system of vectors with all zero coefficients is the zero vector of V. If this is the only way to express the zero vector as a linear combination of $v_1, v_2, \ldots, v_k$ then these vectors are linearly independent. Given a set of vectors that span a space, if any vector w is a linear combination of other vectors (and so the set is not linearly independent), then the span would remain the same if we remove w from the set. Thus, a set of linearly dependent vectors is redundant in the sense that there will be a linearly independent subset which will span the same subspace. Therefore, we are mostly interested in a linearly independent set of vectors that spans a vector space *V*, which we call a basis of *V*. Any set of vectors that spans *V* contains a basis, and any linearly independent set of vectors in *V* can be extended to a basis. It turns out that if we accept the axiom of choice, every vector space has a basis; nevertheless, this basis may be unnatural, and indeed, may not even be constructible. For instance, there exists a basis for the real numbers, considered as a vector space over the rationals, but no explicit basis has been constructed.

Any two bases of a vector space *V* have the same cardinality, which is called the dimension of *V*. The dimension of a vector space is well-defined by the dimension theorem for vector spaces. If a basis of *V* has finite number of elements, *V* is called a finite-dimensional vector space. If *V* is finite-dimensional and *U* is a subspace of *V*, then dim $U \leq$ dim *V*. If U_1 and U_2 are subspaces of *V*, then

$$\dim(U_1 + U_2) = \dim U_1 + \dim U_2 - \dim(U_1 \cap U_2).$$

One often restricts consideration to finite-dimensional vector spaces. A fundamental theorem of linear algebra states that all vector spaces of the same dimension are isomorphic, giving an easy way of characterizing isomorphism.

Matrix Theory

A particular basis $\{v_1, v_2, \ldots, v_n\}$ of *V* allows one to construct a coordinate system in *V*: the vector with coordinates $(a_1, a_2, \ldots, a_n)$ is the linear combination

$$a_1 v_1 + a_2 v_2 + \cdots + a_n v_n.$$

The condition that $v_1, v_2, \ldots, v_n$ span *V* guarantees that each vector *v* can be assigned coordinates, whereas the linear independence of $v_1, v_2, \ldots, v_n$ assures that these coordinates are unique (i.e. there is only one linear combination of the basis vectors that is equal to *v*). In this way, once a basis of a vector space V over F has been chosen, V may be identified with the coordinate n-space F^n. Under this identification, addition and scalar multiplication of vectors in V correspond to addition and scalar multiplication of their coordinate vectors in F^n. Furthermore, if V and W are an n-dimensional and m-dimensional vector space over F, and a basis of V and a basis of W have been fixed, then any linear transformation T: V → W may be encoded by an m × n matrix A with entries in the field F, called the matrix of *T* with respect to these bases. Two matrices that encode the same linear transformation in different bases are called similar. Matrix theory replaces the study of linear transformations, which were defined axiomatically, by the study of matrices, which are concrete objects. This major technique distinguishes linear algebra from theories of other algebraic structures, which usually cannot be parameterized so concretely.

There is an important distinction between the coordinate *n*-space R^n and a general finite-dimensional vector space *V*. While R^n has a standard basis $\{e_1, e_2, \ldots, e_n\}$, a vector space *V* typically does not come equipped with such a basis and many different bases exist (although they all consist of the same number of elements equal to the dimension of *V*).

One major application of the matrix theory is calculation of determinants, a central concept in linear algebra. While determinants could be defined in a basis-free manner, they are usually introduced via a specific representation of the mapping; the value of the determinant does not depend on the specific basis. It turns out that a mapping has an inverse if and only if the determinant has an inverse (every non-zero real or complex number has an inverse). If the determinant is zero, then the nullspace is nontrivial. Determinants have other applications, including a systematic way of seeing if a set of vectors is linearly independent (we write the vectors as the columns of a matrix, and if the determinant of that matrix is zero, the vectors are linearly dependent). Determinants could also be used to solve systems of linear equations, but in real applications, Gaussian elimination is a faster method.

Eigenvalues and Eigenvectors

In general, the action of a linear transformation may be quite complex. Attention to low-dimensional examples gives an indication of the variety of their types. One strategy for a general n-dimensional transformation T is to find "characteristic lines" that are invariant sets under T. If v is a non-zero vector such that Tv is a scalar multiple of v, then the line through 0 and v is an invariant set under T and v is called a characteristic vector or eigenvector. The scalar λ such that $Tv = \lambda v$ is called a characteristic value or eigenvalue of T.

To find an eigenvector or an eigenvalue, we note that

$$\mathrm{Tv} - \lambda \mathrm{v} = (\mathrm{T} - \lambda \mathrm{I})\mathrm{v} = 0,$$

where I is the identity matrix. For there to be nontrivial solutions to that equation, $\det(T - \lambda \mathrm{I}) = 0$. The determinant is a polynomial, and so the eigenvalues are not guaranteed to exist if the field is R. Thus, we often work with an algebraically closed field such as the complex numbers when dealing with eigenvectors and eigenvalues so that an eigenvalue will always exist. It would be particularly nice if given a transformation T taking a vector space V into itself we can find a basis for V consisting of eigenvectors. If such a basis exists, we can easily compute the action of the transformation on any vector: if $v_1, v_2, \ldots, v_n$ are linearly independent eigenvectors of a mapping of n-dimensional spaces T with (not necessarily distinct) eigenvalues $\lambda_1, \lambda_2, \ldots, \lambda_n$, and if $v = a_1 v_1 + \ldots + a_n v_n$, then,

$$\mathrm{T}(\mathrm{v}) = \mathrm{T}(\mathrm{a}_1\mathrm{v}_1) + \cdots + \mathrm{T}(\mathrm{a}_\mathrm{n}\mathrm{v}_\mathrm{n}) = \mathrm{a}_1\mathrm{T}(\mathrm{v}_1) + \cdots + \mathrm{a}_\mathrm{n}\mathrm{T}(\mathrm{v}_\mathrm{n}) = \mathrm{a}_1\lambda_1\mathrm{v}_1 + \cdots + \mathrm{a}_\mathrm{n}\lambda_\mathrm{n}\mathrm{v}_\mathrm{n}.$$

Such a transformation is called a diagonalizable matrix since in the eigenbasis, the transformation is represented by a diagonal matrix. Because operations like matrix multipli-

cation, matrix inversion, and determinant calculation are simple on diagonal matrices, computations involving matrices are much simpler if we can bring the matrix to a diagonal form. Not all matrices are diagonalizable (even over an algebraically closed field).

Inner-product Spaces

Besides these basic concepts, linear algebra also studies vector spaces with additional structure, such as an inner product. The inner product is an example of a bilinear form, and it gives the vector space a geometric structure by allowing for the definition of length and angles. Formally, an *inner product* is a map

$$\langle \cdot,\cdot \rangle : V \times V \to F$$

that satisfies the following three axioms for all vectors u, v, w in V and all scalars a in F:

- Conjugate symmetry:

$$\langle u, v \rangle = \overline{\langle v, u \rangle}.$$

Note that in R, it is symmetric.

- Linearity in the first argument:

$$\langle au, v \rangle = a\langle u, v \rangle.$$

$$\langle u + v, w \rangle = \langle u, w \rangle + \langle v, w \rangle.$$

- Positive-definiteness:

  $\langle v, v \rangle \geq 0$ with equality only for $v = 0$.

We can define the length of a vector v in V by

$$\| v \|^2 = \langle v, v \rangle,$$

and we can prove the Cauchy–Schwarz inequality:

$$|\langle u, v \rangle| \leq \| u \| \cdot \| v \|.$$

In particular, the quantity

$$\frac{|\langle u, v \rangle|}{\| u \| \cdot \| v \|} \leq 1,$$

and so we can call this quantity the cosine of the angle between the two vectors.

Two vectors are orthogonal if $\langle u, v \rangle = 0$. An orthonormal basis is a basis where all basis vectors have length 1 and are orthogonal to each other. Given any finite-dimensional vector space, an orthonormal basis could be found by the Gram–Schmidt procedure. Orthonormal bases are particularly nice to deal with, since if $v = a_1 v_1 + \ldots + a_n v_n$, then $a_i = \langle v, v_i \rangle$.

The inner product facilitates the construction of many useful concepts. For instance, given a transform *T*, we can define its Hermitian conjugate *T** as the linear transform satisfying

$$\langle \mathrm{Tu}, \mathrm{v} \rangle = \langle \mathrm{u}, \mathrm{T}^{*}\mathrm{v} \rangle.$$

If *T* satisfies $TT^* = T^*T$, we call *T* normal. It turns out that normal matrices are precisely the matrices that have an orthonormal system of eigenvectors that span *V*.

Some Main Useful Theorems

- A matrix is invertible, or non-singular, if and only if the linear map represented by the matrix is an isomorphism.
- Any vector space over a field F of dimension n is isomorphic to F^n as a vector space over F.
- Corollary: Any two vector spaces over F of the same finite dimension are isomorphic to each other.
- A linear map is an isomorphism if and only if the determinant is nonzero.

Applications

Because of the ubiquity of vector spaces, linear algebra is used in many fields of mathematics, natural sciences, computer science, and social science. Below are just some examples of applications of linear algebra.

Solution of Linear Systems

Linear algebra provides the formal setting for the linear combination of equations used in the Gaussian method. Suppose the goal is to find and describe the solution(s), if any, of the following system of linear equations:

$$\begin{aligned} 2x + y - z &= 8 && (L_1) \\ -3x - y + 2z &= -11 && (L_2) \\ -2x + y + 2z &= -3 && (L_3) \end{aligned}$$

The Gaussian-elimination algorithm is as follows: eliminate *x* from all equations below L_1, and then eliminate *y* from all equations below L_2. This will put the system into triangular form. Then, using back-substitution, each unknown can be solved for.

In the example, x is eliminated from L_2 by adding $(3/2)L_1$ to L_2. x is then eliminated from L_3 by adding L_1 to L_3. Formally:

$$L_2 + \frac{3}{2}L_1 \rightarrow L_2$$

$$L_3 + L_1 \rightarrow L_3$$

The result is:

$$\begin{aligned}&2x + y - z = 8\\&\frac{1}{2}y + \frac{1}{2}z = 1\\&2y + z = 5\end{aligned}$$

Now y is eliminated from L_3 by adding $-4L_2$ to L_3:

$$L_3 + -4L_2 \rightarrow L_3$$

The result is:

$$\begin{aligned}&2x + y - z - 8\\&\frac{1}{2}y + \frac{1}{2}z = 1\\&-z \ = 1\end{aligned}$$

This result is a system of linear equations in triangular form, and so the first part of the algorithm is complete.

The last part, back-substitution, consists of solving for the known in reverse order. It can thus be seen that

$$z = -1 \quad (L_3)$$

Then, z can be substituted into L_2, which can then be solved to obtain

$$y = 3 \quad (L_2)$$

Next, z and y can be substituted into L_1, which can be solved to obtain

$$x = 2 \quad (L_1)$$

The system is solved.

We can, in general, write any system of linear equations as a matrix equation:

$$Ax = b.$$

The solution of this system is characterized as follows: first, we find a particular solution x_0 of this equation using Gaussian elimination. Then, we compute the solutions of $Ax = 0$; that is, we find the null space N of A. The solution set of this equation is given by $x_0 + N = \{x_0 + n : n \in N\}$. If the number of variables is equal to the number of equations, then we can characterize when the system has a unique solution: since N is trivial if and only if $\det A \neq 0$, the equation has a unique solution if and only if $\det A \neq 0$.

Least-squares Best Fit Line

The least squares method is used to determine the best fit line for a set of data. This line will minimize the sum of the squares of the residuals.

Fourier Series Expansion

Fourier series are a representation of a function f: $[-\pi, \pi] \to R$ as a trigonometric series:

$$f(x) = \frac{a_0}{2} + \sum_{n=1}^{\infty} [a_n \cos(nx) + b_n \sin(nx)].$$

This series expansion is extremely useful in solving partial differential equations. In this article, we will not be concerned with convergence issues; it is nice to note that all Lipschitz-continuous functions have a converging Fourier series expansion, and nice enough discontinuous functions have a Fourier series that converges to the function value at most points.

The space of all functions that can be represented by a Fourier series form a vector space (technically speaking, we call functions that have the same Fourier series expansion the "same" function, since two different discontinuous functions might have the same Fourier series). Moreover, this space is also an inner product space with the inner product

$$\langle f, g \rangle = \frac{1}{\pi} \int_{-\pi}^{\pi} f(x) g(x) dx.$$

The functions $g_n(x) = \sin(nx)$ for $n > 0$ and $h_n(x) = \cos(nx)$ for $n \geq 0$ are an orthonormal basis for the space of Fourier-expandable functions. We can thus use the tools of linear algebra to find the expansion of any function in this space in terms of these basis functions. For instance, to find the coefficient a_k, we take the inner product with h_k:

$$\langle f, h_k \rangle = \frac{a_0}{2}\langle h_0, h_k \rangle + \sum_{n=1}^{\infty} [a_n \langle h_n, h_k \rangle + b_n \langle g_n, h_k \rangle],$$

and by orthonormality, $\langle f, h_k \rangle = a_k$; that is,

$$a_k = \frac{1}{\pi}\int_{-\pi}^{\pi} f(x)\cos(kx)dx.$$

Quantum Mechanics

Quantum mechanics is highly inspired by notions in linear algebra. In quantum mechanics, the physical state of a particle is represented by a vector, and observables (such as momentum, energy, and angular momentum) are represented by linear operators on the underlying vector space. More concretely, the wave function of a particle describes its physical state and lies in the vector space L^2 (the functions $\varphi: R^3 \to C$ such that $\int_{-\infty}^{\infty}\int_{-\infty}^{\infty}\int_{-\infty}^{\infty} |\phi|^2 \, dxdydz$ is finite), and it evolves according to the Schrödinger equation. Energy is represented as the operator $H = -\frac{\hbar^2}{2m}\nabla^2 + V(x,y,z)$, where V is the potential energy. H is also known as the Hamiltonian operator. The eigenvalues of H represents the possible energies that can be observed. Given a particle in some state φ, we can expand φ into a linear combination of eigenstates of H. The component of H in each eigenstate determines the probability of measuring the corresponding eigenvalue, and the measurement forces the particle to assume that eigenstate (wave function collapse).

Geometric Introduction

Many of the principles and techniques of linear algebra can be seen in the geometry of lines in a real two dimensional plane E. When formulated using vectors and matrices the geometry of points and lines in the plane can be extended to the geometry of points and hyperplanes in high-dimensional spaces.

Point coordinates in the plane E are ordered pairs of real numbers, (x,y), and a line is defined as the set of points (x,y) that satisfy the linear equation

$$\lambda : ax + by + c = 0,$$

where a, b and c are not all zero. Then,

$$\lambda : [a \quad b \quad c]\begin{Bmatrix} x \\ y \\ 1 \end{Bmatrix} = 0,$$

or

$$Ax = 0,$$

where $x = (x, y, 1)$ is the 3×1 set of homogeneous coordinates associated with the point (x, y).

Homogeneous coordinates identify the plane E with the $z = 1$ plane in three dimensional space. The x–y coordinates in E are obtained from homogeneous coordinates $y = (y_1, y_2, y_3)$ by dividing by the third component (if it is nonzero) to obtain $y = (y_1/y_3, y_2/y_3, 1)$.

The linear equation, λ, has the important property, that if x_1 and x_2 are homogeneous coordinates of points on the line, then the point $\alpha x_1 + \beta x_2$ is also on the line, for any real α and β.

Now consider the equations of the two lines λ_1 and λ_2,

$$\lambda_1 : a_1 x + b_1 y + c_1 = 0, \quad \lambda_2 : a_2 x + b_2 y + c_2 = 0,$$

which forms a system of linear equations. The intersection of these two lines is defined by $x = (x, y, 1)$ that satisfy the matrix equation,

$$\lambda_{1,2} : \begin{bmatrix} a_1 & b_1 & c_1 \\ a_2 & b_2 & c_2 \end{bmatrix} \begin{Bmatrix} x \\ y \\ 1 \end{Bmatrix} = \begin{Bmatrix} 0 \\ 0 \end{Bmatrix},$$

or using homogeneous coordinates,

$$Bx = 0.$$

The point of intersection of these two lines is the unique non-zero solution of these equations. In homogeneous coordinates, the solutions are multiples of the following solution:

$$x_1 = \begin{vmatrix} b_1 & c_1 \\ b_2 & c_2 \end{vmatrix}, x_2 = -\begin{vmatrix} a_1 & c_1 \\ a_2 & c_2 \end{vmatrix}, x_3 = \begin{vmatrix} a_1 & b_1 \\ a_2 & b_2 \end{vmatrix}$$

if the rows of B are linearly independent (i.e., λ_1 and λ_2 represent distinct lines). Divide through by x_3 to get Cramer's rule for the solution of a set of two linear equations in two unknowns. Notice that this yields a point in the $z = 1$ plane only when the 2×2 submatrix associated with x_3 has a non-zero determinant.

It is interesting to consider the case of three lines, λ_1, λ_2 and λ_3, which yield the matrix equation,

$$\lambda_{1,2,3}:\begin{bmatrix} a_1 & b_1 & c_1 \\ a_2 & b_2 & c_2 \\ a_3 & b_3 & c_3 \end{bmatrix}\begin{Bmatrix} x \\ y \\ 1 \end{Bmatrix}=\begin{Bmatrix} 0 \\ 0 \\ 0 \end{Bmatrix}.$$

which in homogeneous form yields,

$$Cx = 0.$$

Clearly, this equation has the solution x = (0,0,0), which is not a point on the $z = 1$ plane *E*. For a solution to exist in the plane *E*, the coefficient matrix *C* must have rank 2, which means its determinant must be zero. Another way to say this is that the columns of the matrix must be linearly dependent.

Introduction to Linear Transformations

Another way to approach linear algebra is to consider linear functions on the two dimensional real plane E=R². Here R denotes the set of real numbers. Let x=(x, y) be an arbitrary vector in *E* and consider the linear function λ: *E*→R, given by

$$\lambda:\begin{bmatrix} a & b \end{bmatrix}\begin{Bmatrix} x \\ y \end{Bmatrix}=c,$$

or

$$Ax = c.$$

This transformation has the important property that if Ay=d, then

$$A(\alpha x+\beta y)=\alpha Ax+\beta Ay=\alpha c+\beta d.$$

This shows that the sum of vectors in *E* map to the sum of their images in R. This is the defining characteristic of a linear map, or linear transformation. For this case, where the image space is a real number the map is called a linear functional.

Consider the linear functional a little more carefully. Let i=(1,0) and j =(0,1) be the natural basis vectors on *E*, so that x=xi+yj. It is now possible to see that

$$Ax = A(xi+yj)=xAi+yAj=\begin{bmatrix} Ai & Aj \end{bmatrix}\begin{Bmatrix} x \\ y \end{Bmatrix}=\begin{bmatrix} a & b \end{bmatrix}\begin{Bmatrix} x \\ y \end{Bmatrix}=c.$$

Thus, the columns of the matrix A are the image of the basis vectors of *E* in R.

This is true for any pair of vectors used to define coordinates in *E*. Suppose we select a non-orthogonal non-unit vector basis v and w to define coordinates of vectors in *E*. This means a vector x has coordinates (α,β), such that x=αv+βw. Then, we have the linear functional

$$\lambda : \mathrm{Ax} = \begin{bmatrix} \mathrm{Av} & \mathrm{Aw} \end{bmatrix} \begin{Bmatrix} \alpha \\ \beta \end{Bmatrix} = \begin{bmatrix} \mathrm{d} & \mathrm{e} \end{bmatrix} \begin{Bmatrix} \alpha \\ \beta \end{Bmatrix} = \mathrm{c},$$

where Av=d and Aw=e are the images of the basis vectors v and w. This is written in matrix form as

$$\begin{bmatrix} \mathrm{a} & \mathrm{b} \end{bmatrix} \begin{bmatrix} \mathrm{v}_1 & \mathrm{w}_1 \\ \mathrm{v}_2 & \mathrm{w}_2 \end{bmatrix} = \begin{bmatrix} \mathrm{d} & \mathrm{e} \end{bmatrix}.$$

Coordinates Relative to a Basis

This leads to the question of how to determine the coordinates of a vector x relative to a general basis v and w in *E*. Assume that we know the coordinates of the vectors, x, v and w in the natural basis i=(1,0) and j =(0,1). Our goal is two find the real numbers α, β, so that x=αv+βw, that is

$$\begin{Bmatrix} \mathrm{x} \\ \mathrm{y} \end{Bmatrix} = \begin{bmatrix} \mathrm{v}_1 & \mathrm{w}_1 \\ \mathrm{v}_2 & \mathrm{w}_2 \end{bmatrix} \begin{Bmatrix} \alpha \\ \beta \end{Bmatrix}.$$

To solve this equation for α, β, we compute the linear coordinate functionals σ and τ for the basis v, w, which are given by,

$$\sigma = \begin{bmatrix} \sigma_1 & \sigma_2 \end{bmatrix} = \frac{1}{\mathrm{v}_1\mathrm{w}_2 - \mathrm{v}_2\mathrm{w}_1} \begin{bmatrix} \mathrm{w}_2 & -\mathrm{w}_1 \end{bmatrix}, \tau = \begin{bmatrix} \tau_1 & \tau_2 \end{bmatrix} = \frac{1}{\mathrm{v}_1\mathrm{w}_2 - \mathrm{v}_2\mathrm{w}_1} \begin{bmatrix} -\mathrm{v}_2 & \mathrm{v}_1 \end{bmatrix},$$

The functionals σ and τ compute the components of x along the basis vectors v and w, respectively, that is,

$$\sigma \mathrm{x} = \alpha, \tau \mathrm{x} = \beta,$$

which can be written in matrix form as

$$\begin{bmatrix} \sigma_1 & \sigma_2 \\ \tau_1 & \tau_2 \end{bmatrix} \begin{Bmatrix} \mathrm{x} \\ \mathrm{y} \end{Bmatrix} = \begin{Bmatrix} \alpha \\ \beta \end{Bmatrix}.$$

These coordinate functionals have the properties,

$$\sigma \mathrm{v} = 1, \sigma \mathrm{w} = 0, \tau \mathrm{w} = 1, \tau \mathrm{v} = 0.$$

These equations can be assembled into the single matrix equation,

$$\begin{bmatrix} \sigma_1 & \sigma_2 \\ \tau_1 & \tau_2 \end{bmatrix} \begin{bmatrix} v_1 & w_1 \\ v_2 & w_2 \end{bmatrix} = \begin{bmatrix} 1 & 0 \\ 0 & 1 \end{bmatrix}.$$

Thus, the matrix formed by the coordinate linear functionals is the inverse of the matrix formed by the basis vectors.

Inverse Image

The set of points in the plane *E* that map to the same image in R under the linear functional λ define a line in E. This line is the image of the inverse map, λ^{-1}: R→*E*. This inverse image is the set of the points x=(x, y) that solve the equation,

$$Ax = [a \quad b]\begin{Bmatrix} x \\ y \end{Bmatrix} = c.$$

Notice that a linear functional operates on known values for x=(x, y) to compute a value *c* in R, while the inverse image seeks the values for x=(x, y) that yield a specific value *c*.

In order to solve the equation, we first recognize that only one of the two unknowns (x,y) can be determined, so we select y to be determined, and rearrange the equation

$$by = c - ax.$$

Solve for y and obtain the inverse image as the set of points,

$$x(t) = \begin{Bmatrix} 0 \\ c/b \end{Bmatrix} + t\begin{Bmatrix} 1 \\ -a/b \end{Bmatrix} = p + th.$$

For convenience the free parameter x has been relabeled t.

The vector p defines the intersection of the line with the y-axis, known as the y-intercept. The vector h satisfies the homogeneous equation,

$$Ah = [a \quad b]\begin{Bmatrix} 1 \\ -a/b \end{Bmatrix} = 0.$$

Notice that if h is a solution to this homogeneous equation, then th is also a solution.

The set of points of a linear functional that map to zero define the kernel of the linear functional. The line can be considered to be the set of points h in the kernel translated by the vector p.

Generalizations and Related Topics

Since linear algebra is a successful theory, its methods have been developed and generalized in other parts of mathematics. In module theory, one replaces the field of scalars by a ring. The concepts of linear independence, span, basis, and dimension (which is called rank in module theory) still make sense. Nevertheless, many theorems from linear algebra become false in module theory. For instance, not all modules have a basis (those that do are called free modules), the rank of a free module is not necessarily unique, not every linearly independent subset of a module can be extended to form a basis, and not every subset of a module that spans the space contains a basis.

In multilinear algebra, one considers multivariable linear transformations, that is, mappings that are linear in each of a number of different variables. This line of inquiry naturally leads to the idea of the dual space, the vector space V^* consisting of linear maps $f: V \rightarrow F$ where F is the field of scalars. Multilinear maps $T: V^n \rightarrow F$ can be described via tensor products of elements of V^*.

If, in addition to vector addition and scalar multiplication, there is a bilinear vector product $V \times V \rightarrow V$, the vector space is called an algebra; for instance, associative algebras are algebras with an associate vector product (like the algebra of square matrices, or the algebra of polynomials).

Functional analysis mixes the methods of linear algebra with those of mathematical analysis and studies various function spaces, such as L^p spaces.

Representation theory studies the actions of algebraic objects on vector spaces by representing these objects as matrices. It is interested in all the ways that this is possible, and it does so by finding subspaces invariant under all transformations of the algebra. The concept of eigenvalues and eigenvectors is especially important.

Algebraic geometry considers the solutions of systems of polynomial equations.

There are several related topics in the field of Computer Programming that utilizes much of the techniques and theorems Linear Algebra encompasses and refers to.

Basis (Linear Algebra)

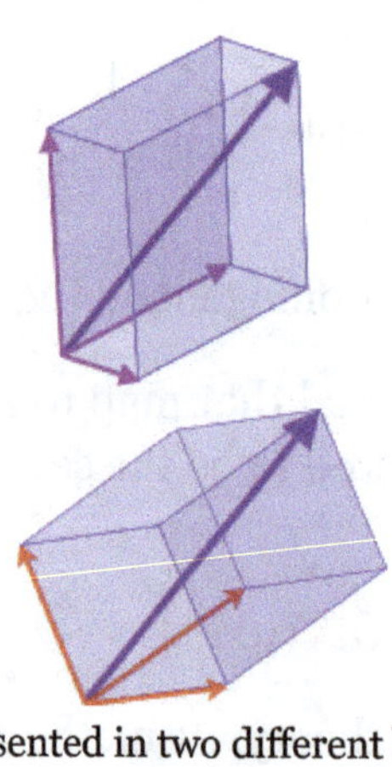

The same vector can be represented in two different bases (purple and red arrows).

In mathematics, a set of elements (vectors) in a vector space V is called a basis, or a set of basis vectors, if the vectors are linearly independent and every vector in the vector space is a linear combination of this set. In more general terms, a basis is a linearly independent spanning set.

Given a basis of a vector space V, every element of V can be expressed uniquely as a linear combination of basis vectors, whose coefficients are referred to as vector coordinates or components. A vector space can have several distinct sets of basis vectors; however each such set has the same number of elements, with this number being the dimension of the vector space.

Definition

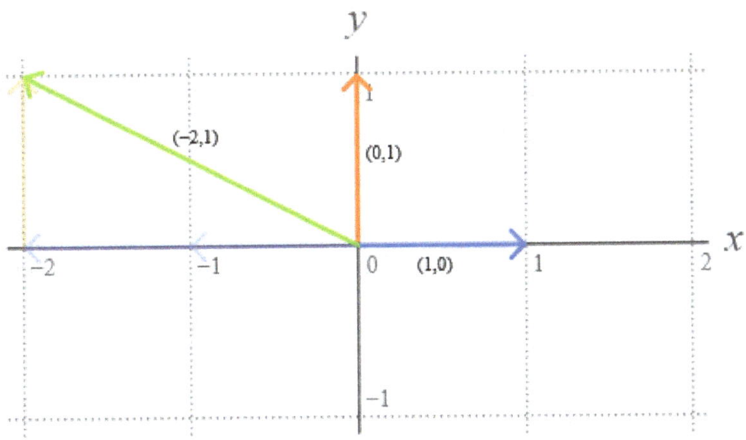

This picture illustrates the standard basis in R^2. The blue and orange vectors are the elements of the basis; the green vector can be given in terms of the basis vectors, and so is linearly dependent upon them.

A basis B of a vector space V over a field F is a linearly independent subset of V that spans V.

In more detail, suppose that $B = \{ v_1, \ldots, v_n \}$ is a finite subset of a vector space V over a field F (such as the real or complex numbers R or C). Then B is a basis if it satisfies the following conditions:

- the *linear independence* property,

  for all $a_1, \ldots, a_n \in$ F, if $a_1v_1 + \ldots + a_nv_n = 0$, then necessarily $a_1 = \ldots = a_n = 0$; and

- the *spanning* property,

  for every (vector) x in V it is possible to choose $a_1, \ldots, a_n \in$ F such that $x = a_1v_1 + \ldots + a_nv_n$.

The numbers a_i are called the coordinates of the vector x with respect to the basis B, and by the first property they are uniquely determined.

A vector space that has a finite basis is called finite-dimensional. To deal with infinite-dimensional spaces, we must generalize the above definition to include infinite basis sets. We therefore say that a set (finite or infinite) $B \subset V$ is a basis, if

- every finite subset $B_0 \subseteq B$ obeys the independence property shown above; and
- for every x in V it is possible to choose $a_1, \ldots, a_n \in$ F and $v_1, \ldots, v_n \in B$ such that $x = a_1v_1 + \ldots + a_nv_n$.

The sums in the above definition are all finite because without additional structure the axioms of a vector space do not permit us to meaningfully speak about an infinite sum of vectors. Settings that permit infinite linear combinations allow alternative definitions of the basis concept.

It is often convenient to list the basis vectors in a specific *order*, for example, when considering the transformation matrix of a linear map with respect to a basis. We then speak of an ordered basis, which we define to be a sequence (rather than a set) of linearly independent vectors that span V.

Properties

Again, B denotes a subset of a vector space V. Then, B is a basis if and only if any of the following equivalent conditions are met:

- B is a minimal generating set of V, i.e., it is a generating set and no proper subset of B is also a generating set.
- B is a maximal set of linearly independent vectors, i.e., it is a linearly independent set but no other linearly independent set contains it as a proper subset.
- Every vector in V can be expressed as a linear combination of vectors in B in a unique way. If the basis is ordered then the coefficients in this linear combination provide *coordinates* of the vector relative to the basis.

Every vector space has a basis. The proof of this requires the axiom of choice. All bases of a vector space have the same cardinality (number of elements), called the dimension of the vector space. This result is known as the dimension theorem, and requires the ultrafilter lemma, a strictly weaker form of the axiom of choice.

Also many vector sets can be attributed a standard basis which comprises both spanning and linearly independent vectors.

Standard bases for example:

In R^n, $\{e_1, \ldots, e_n\}$, where e_i is the ith column of the identity matrix.

In P_2, where P_2 is the set of all polynomials of degree at most 2, $\{1, x, x^2\}$ is the standard basis.

In M_{22}, $\{M_{1,1}, M_{1,2}, M_{2,1}, M_{2,2}\}$, where M_{22} is the set of all 2×2 matrices. and $M_{m,n}$ is the 2×2 matrix with a 1 in the m,n position and zeros everywhere else.

Constructing a Basis

We use the following algorithm,

Step 1: If V = {0}, then the empty set is a basis. Otherwise, let $v_1 \in V \setminus \{0\}$. Then $\{v_1\}$ is linearly independent.

Step 2: If $V = LS(v_1)$, we have got a basis of V . Else, pick $v_2 \in V \setminus LS(v_1)$. Then, $\{v_1, v_2\}$ is linearly independent.

Step 3: Either $V = LS(v_1, v_2, \ldots, v_i)$ or $LS(v_1, v_2, \ldots, v_i)$ is not equal to V . In the first case, $\{v_1, v_2, \ldots, v_i\}$ is a basis of V . Else, pick $v_i+1 \in V \setminus LS(v_1, v_2, \ldots, v_i)$. Then, the set $\{v_1, v_2, \ldots, v_i+1\}$ is linearly independent.

This process will finally end as V is a finite dimensional vector space.

Change of Basis

Given a vector space V over a field F and suppose that $\{v_1, ..., v_n\}$ and $\{\alpha_1, ..., \alpha_n\}$ are two bases for V. By definition, if ξ is a vector in V then $\xi = x_1\alpha_1 + ... + x_n\alpha_n$ for a unique choice of scalars $x_1, ..., x_n$ in F called the *coordinates of* ξ *relative to the ordered basis* $\{\alpha_1, ..., \alpha_n\}$. The vector $x = (x_1, ..., x_n)$ in F^n is called the *coordinate tuple of* ξ (relative to this basis). The unique linear map $\varphi : F^n \rightarrow V$ with $\varphi(v_j) = \alpha_j$ for $j = 1, ..., n$ is called the *coordinate isomorphism* for V and the basis $\{\alpha_1, ..., \alpha_n\}$. Thus $\varphi(x) = \xi$ if and only if $\xi = x_1\alpha_1 + ... + x_n\alpha_n$.

A set of vectors can be represented by a matrix of which each column consists of the components of the corresponding vector of the set. As a basis is a set of vectors, a basis can be given by a matrix of this kind. The change of basis of any object of the space is related to this matrix. For example, coordinate tuples change with its inverse.

Examples

- Consider R^2, the vector space of all coordinates (a, b) where both a and b are real numbers. Then a very natural and simple basis is simply the vectors $e_1 = (1,0)$ and $e_2 = (0,1)$: suppose that v = (a, b) is a vector in R^2, then v = a(1,0) + b(0,1). But any two linearly independent vectors, like (1,1) and (−1,2), will also form a basis of R^2.

- More generally, the vectors $e_1, e_2, ..., e_n$ are linearly independent and generate R^n. Therefore, they form a basis for R^n and the dimension of R^n is n. This basis is called the standard basis.

- Let V be the real vector space generated by the functions e^t and e^{2t}. These two functions are linearly independent, so they form a basis for V.
- Let R[x] denote the vector space of real polynomials; then $(1, x, x^2, \ldots)$ is a basis of R[x]. The dimension of R[x] is therefore equal to aleph-0.

Extending to a Basis

Let S be a subset of a vector space V. To extend S to a basis means to find a basis B that contains S as a subset. This can be done if and only if S is linearly independent. Almost always, there is more than one such B, except in rather special circumstances (i.e. S is already a basis, or S is empty and V has two elements).

A similar question is when does a subset S contain a basis. This occurs if and only if S spans V. In this case, S will usually contain several different bases.

Example of Alternative Proofs

Often, a mathematical result can be proven in more than one way. Here, using three different proofs, we show that the vectors (1,1) and (−1,2) form a basis for R^2.

From the Definition of *Basis*

We have to prove that these two vectors are linearly independent and that they generate R^2.

Part I: If two vectors v and w are linearly independent, then $av + bw = 0$ (a and b scalars) implies $a = 0, b = 0$.

To prove that they are linearly independent, suppose that there are numbers a, b such that:

$$a(1,1) + b(-1,2) = (0,0)$$

(i.e., they are linearly dependent). Then:

$(a - b, a + 2b) = (0,0)$ and $a - b = 0$ and $a + 2b = 0$.

Subtracting the first equation from the second, we obtain:

$3b = 0$ so $b = 0$.

Adding this equation to the first equation then:

$$a = 0.$$

Hence we have linear independence.

Part II: To prove that these two vectors generate R^2, we have to let (a, b) be an arbitrary element of R^2, and show that there exist numbers r, s ∈ R such that:

$$r(1,1)+s(-1,2)=(a,b).$$

Then we have to solve the equations:

$$r-s=a$$

$$r+2s=b.$$

Subtracting the first equation from the second, we get:

$3s=b-a$, and then

$s=(b-a)/3$ and finally

$$r=s+a=((b-a)/3)+a=(b+2a)/3.$$

By the Dimension Theorem

Since (−1,2) is clearly not a multiple of (1,1) and since (1,1) is not the zero vector, these two vectors are linearly independent. Since the dimension of R^2 is 2, the two vectors already form a basis of R^2 without needing any extension.

By the Invertible Matrix Theorem

Simply compute the determinant

$$\det\begin{bmatrix}1 & -1\\ 1 & 2\end{bmatrix}=3\neq 0.$$

Since the above matrix has a nonzero determinant, its columns form a basis of R^2.

Ordered Bases and Coordinates

A basis is just a linearly independent *set* of vectors with or without a given ordering. For many purposes it is convenient to work with an ordered basis. For example, when working with a coordinate representation of a vector it is customary to speak of the "first" or "second" coordinate, which makes sense only if an ordering is specified for the basis. For finite-dimensional vector spaces one typically indexes a basis $\{v_i\}$ by the first n integers. An ordered basis is also called a frame.

Suppose *V* is an *n*-dimensional vector space over a field F. A choice of an ordered basis for V is equivalent to a choice of a linear isomorphism φ from the coordinate space F^n to *V*.

Proof. The proof makes use of the fact that the standard basis of F^n is an ordered basis. Suppose first that

$$\varphi : \mathbf{F}^n \to V$$

is a linear isomorphism. Define an ordered basis $\{v_i\}$ for V by

$$v_i = \varphi(\mathbf{e}_i) \text{ for } 1 \le i \le n$$

where $\{e_i\}$ is the standard basis for F^n.

Conversely, given an ordered basis, consider the map defined by

$$\varphi(x) = x_1v_1 + x_2v_2 + \ldots + x_nv_n,$$

where $x = x_1e_1 + x_2e_2 + \ldots + x_ne_n$ is an element of F^n. It is not hard to check that φ is a linear isomorphism.

These two constructions are clearly inverse to each other. Thus ordered bases for V are in 1-1 correspondence with linear isomorphisms $F^n \to V$.

The inverse of the linear isomorphism φ determined by an ordered basis $\{v_i\}$ equips V with *coordinates*: if, for a vector $v \in V$, $\varphi^{-1}(v) = (a_1, a_2,\ldots,a_n) \in F^n$, then the components $a_j = a_j(v)$ are the coordinates of v in the sense that $v = a_1(v)\, v_1 + a_2(v)\, v_2 + \ldots + a_n(v)\, v_n$.

The maps sending a vector v to the components $a_j(v)$ are linear maps from V to F, because of φ^{-1} is linear. Hence they are linear functionals. They form a basis for the dual space of V, called the dual basis.

Related Notions

Analysis

In the context of infinite-dimensional vector spaces over the real or complex numbers, the term *Hamel basis* (named after Georg Hamel) or *algebraic basis* can be used to refer to a basis as defined in this article. This is to make a distinction with other notions of "basis" that exist when infinite-dimensional vector spaces are endowed with extra structure. The most important alternatives are orthogonal bases on Hilbert spaces, Schauder bases and Markushevich bases on normed linear spaces. The term *Hamel* ***basis*** is also commonly used to mean a basis for the real numbers R as a vector space over the field Q of rational numbers. (In this case, the dimension of R over Q is uncountable, specifically the continuum, the cardinal number $2^{\aleph_0} = \mathfrak{c}$.) A consequence of the existence of a basis for R over Q is the existence of non-linear additive functions on R.

The common feature of the other notions is that they permit the taking of infinite linear combinations of the basis vectors in order to generate the space. This, of course, requires that infinite sums are meaningfully defined on these spaces, as is the case for topological vector spaces – a large class of vector spaces including e.g. Hilbert spaces, Banach spaces or Fréchet spaces.

The preference of other types of bases for infinite-dimensional spaces is justified by the fact that the Hamel basis becomes "too big" in Banach spaces: If X is an infinite-dimensional normed vector space which is complete (i.e. X is a Banach space), then any Hamel basis of X is necessarily uncountable. This is a consequence of the Baire category theorem. The completeness as well as infinite dimension are crucial assumptions in the previous claim. Indeed, finite-dimensional spaces have by definition finite bases and there are infinite-dimensional (*non-complete*) normed spaces which have countable Hamel bases. Consider c_{00}, the space of the sequences $x = (x_n)$ of real numbers which have only finitely many non-zero elements, with the norm $\| x \| = \sup_n | x_n |$. Its standard basis, consisting of the sequences having only one non-zero element, which is equal to 1, is a countable Hamel basis.

Example

In the study of Fourier series, one learns that the functions {1} ∪ { sin(nx), cos(nx) : n = 1, 2, 3, ... } are an "orthogonal basis" of the (real or complex) vector space of all (real or complex valued) functions on the interval [0, 2π] that are square-integrable on this interval, i.e., functions f satisfying

$$\int_0^{2\pi} |f(x)|^2 dx < \infty.$$

The functions {1} ∪ { sin(nx), cos(nx) : n = 1, 2, 3, ... } are linearly independent, and every function f that is square-integrable on [0, 2π] is an "infinite linear combination" of them, in the sense that

$$\lim_{n\to\infty} \int_0^{2\pi} \left| a_0 + \sum_{k=1}^{n} \big(a_k \cos(kx) + b_k \sin(kx)\big) - f(x) \right|^2 dx = 0$$

for suitable (real or complex) coefficients a_k, b_k. But many square-integrable functions cannot be represented as *finite* linear combinations of these basis functions, which therefore *do not* comprise a Hamel basis. Every Hamel basis of this space is much bigger than this merely countably infinite set of functions. Hamel bases of spaces of this kind are typically not useful, whereas orthonormal bases of these spaces are essential in Fourier analysis.

Geometry

The geometric notions of an affine space, projective space, convex set, and cone have related notions of basis. An affine basis for an n-dimensional affine space is $n+1$ points in general linear position. A projective basis is $n+2$ points in general position, in a projective space of dimension n. A convex basis of a polytope is the set of the vertices of its convex hull. A cone basis consists of one point by edge of a polygonal cone.

Random Basis

Consider a probability distribution in R^n, which admits probability density function (for example, the equidistribution in a n-dimensional ball with respect to Lebesgue measure). Randomly and independently chosen n vectors form a basis with probability one because n linearly dependent vectors $x_1,..., x_n$ in R^n should satisfy the equation $det[x_1,..., x_n]=0$ (zero determinant of the matrix with columns x_i), and the set of zeros of a non-trivial polynomial has zero measure. This simple observation gave rise to various techniques of approximation in random bases.

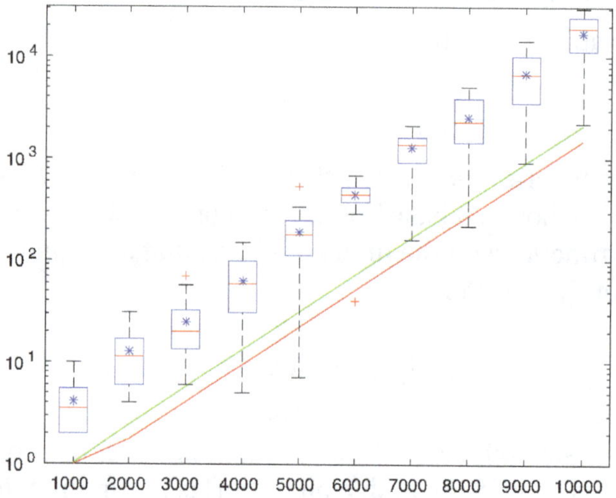

Empirical distribution of lengths N of pairwise almost orthogonal chains of vectors that are independently randomly sampled from the n-dimensional cube $[-1,1]^n$ as a function of dimension, n. Boxplots show the second and third quartiles of this data for each n, red bars correspond to the medians, and blue stars indicate means. Red curve shows theoretical bound given by Eq. (1) and green curve shows a refined estimate.

It is difficult to check numerically the linear dependence or exact orthogonality. Therefore, the notion of ε-orthogonality is used. For spaces with inner product, x is ε-orthogonal to y if $|\langle x,y\rangle|/(\|x\|\ \|y\|)<\epsilon$ (that is, cosine of the angle between x and y is less than ε).

In high dimensions, two independent random vectors are with high probability almost orthogonal, and the number of independent random vectors, which all are with given high probability pairwise almost orthogonal, grows exponentially with dimension. More precisely, consider equidistribution in n-dimensional ball. Choose N independent random vectors from a ball (they are independent and identically distributed). Let θ be a small positive number. Then for

$$N \leq e^{\frac{\epsilon^2 n}{4}} [-\ln(1-\theta)]^{\frac{1}{2}}$$

N random vectors are all pairwise ε-orthogonal with probability 1-θ. This *N* growth exponentially with dimension *n* and $N \gg n$ for sufficiently big *n*. This property of random bases is a manifestation of the so-called *measure concentration phenomenon.*

The figure illustrates distribution of lengths N of pairwise almost orthogonal chains of vectors that are independently randomly sampled from the *n*-dimensional cube $[-1,1]^n$ as a function of dimension, *n*. A point is first randomly selected in the cube. The second point is randomly chosen in the same cube. If the angle between the vectors was within $\pi/2 \pm 0.037\pi/2$ then the vector was retained. At the next step a new vector is generated in the same hypercube, and its angles with the previously generated vectors are evaluated. If these angles are within $\pi/2 \pm 0.037\pi/2$ then the vector is retained. The process is repeated until the chain of almost orthogonality breaks, and the number of such pairwise almost orthogonal vectors (length of the chain) is recorded. For each *n*, 20 pairwise almost orthogonal chains where constructed numerically for each dimension. Distribution of the length of these chains is presented.

Proof that Every Vector Space has a Basis

Let V be any vector space over some field F. Let X be the set of all linearly independent subsets of V.

The set X is nonempty since the empty set is an independent subset of V, and it is partially ordered by inclusion, which is denoted, as usual, by ⊆.

Let Y be a subset of X that is totally ordered by ⊆, and let L_Y be the union of all the elements of Y (which are themselves certain subsets of V).

Since (Y, ⊆) is totally ordered, every finite subset of L_Y is a subset of an element of Y, which is a linearly independent subset of V, and hence every finite subset of L_Y is linearly independent. Thus L_Y is linearly independent, so L_Y is an element of X. Therefore L_Y is an upper bound for Y in (X, ⊆): it is an element of X, that contains every element Y.

As X is nonempty, and every totally ordered subset of (X, ⊆) has an upper bound, Zorn's lemma asserts that X has a maximal element. In other words, there exists some element L_{max} of X satisfying the condition that whenever $L_{max} \subseteq L$ for some element L of X, then $L = L_{max}$.

It remains to prove that L_{max} is a basis of V. Since L_{max} belongs to X, we already know that L_{max} is a linearly independent subset of V.

If L_{max} would not span V, there would exist some vector w of V that cannot be expressed as a linear combination of elements of L_{max} (with coefficients in the field F). In particular, w cannot be an element of L_{max}. Let $L_w = L_{max} \cup \{w\}$. This set is an element of X, that is, it is a linearly independent subset of V (because w is not in the span of L_{max}, and L_{max} is independent). As $L_{max} \subseteq L_w$, and $L_{max} \neq L_w$ (because L_w contains the vector w that is not contained in L_{max}), this contradicts the maximality of L_{max}. Thus this shows that L_{max} spans V.

Hence L_{max} is linearly independent and spans V. It is thus a basis of V, and this proves that every vector space has a basis.

This proof relies on Zorn's lemma, which is equivalent to the axiom of choice. Conversely, it may be proved that if every vector space has a basis, then the axiom of choice is true; thus the two assertions are equivalent.

Vector Space

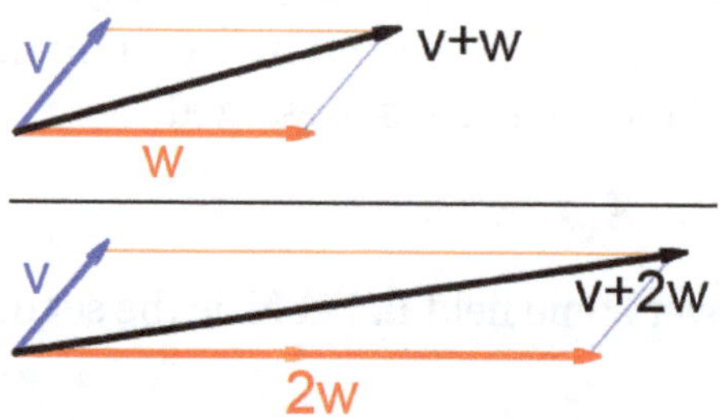

Vector addition and scalar multiplication: a vector v (blue) is added to another vector w (red, upper illustration). Below, w is stretched by a factor of 2, yielding the sum v + 2w.

A vector space (also called a linear space) is a collection of objects called vectors, which may be added together and multiplied ("scaled") by numbers, called scalars in this context. Scalars are often taken to be real numbers, but there are also vector spaces with scalar multiplication by complex numbers, rational numbers, or generally any field. The operations of vector addition and scalar multiplication must satisfy certain requirements, called *axioms*.

Euclidean vectors are an example of a vector space. They represent physical quantities such as forces: any two forces (of the same type) can be added to yield a third, and the multiplication of a force vector by a real multiplier is another force vector. In the same vein, but in a more geometric sense, vectors representing displacements in the plane or in three-dimensional space also form vector spaces. Vectors in vector spaces do not necessarily have to be arrow-like objects as they appear in the mentioned examples: vectors are regarded as abstract mathematical objects with particular properties, which in some cases can be visualized as arrows.

Vector spaces are the subject of linear algebra and are well characterized by their dimension, which, roughly speaking, specifies the number of independent directions in the space. Infinite-dimensional vector spaces arise naturally in mathematical analysis, as function spaces, whose vectors are functions. These vector spaces are generally endowed with additional structure, which may be a topology, allowing the consideration of issues of proximity and continuity. Among these topologies, those that are defined by a norm or inner product are more commonly used, as having a notion of distance between two vectors. This is particularly the case of Banach spaces and Hilbert spaces, which are fundamental in mathematical analysis.

Historically, the first ideas leading to vector spaces can be traced back as far as the 17th century's analytic geometry, matrices, systems of linear equations, and Euclidean vectors. The modern, more abstract treatment, first formulated by Giuseppe Peano in 1888, encompasses more general objects than Euclidean space, but much of the theory can be seen as an extension of classical geometric ideas like lines, planes and their higher-dimensional analogs.

Today, vector spaces are applied throughout mathematics, science and engineering. They are the appropriate linear-algebraic notion to deal with systems of linear equations; offer a framework for Fourier expansion, which is employed in image compression routines; or provide an environment that can be used for solution techniques for partial differential equations. Furthermore, vector spaces furnish an abstract, coordinate-free way of dealing with geometrical and physical objects such as tensors. This in turn allows the examination of local properties of manifolds by linearization techniques. Vector spaces may be generalized in several ways, leading to more advanced notions in geometry and abstract algebra.

Definition

The concept of vector space will first be explained by describing two particular examples:

First Example: Arrows in the Plane

The first example of a vector space consists of arrows in a fixed plane, starting at one fixed point. This is used in physics to describe forces or velocities. Given any two such arrows, v and w, the parallelogram spanned by these two arrows contains one diagonal arrow that starts at the origin, too. This new arrow is called the sum of the two arrows and is denoted v + w. In the special case of two arrows on the same line, their sum is the arrow on this line whose length is the sum or the difference of the lengths, depending on whether the arrows have the same direction. Another operation that can be done with arrows is scaling: given any positive real number a, the arrow that has the same direction as v, but is dilated or shrunk by multiplying its length by a, is called *multiplication* of v by a. It is denoted av. When a is negative, av is defined as the arrow pointing in the opposite direction, instead.

The following shows a few examples: if $a = 2$, the resulting vector aw has the same direction as w, but is stretched to the double length of w (right image below). Equivalently 2w is the sum w + w. Moreover, (−1)v = −v has the opposite direction and the same length as v (blue vector pointing down in the right image).

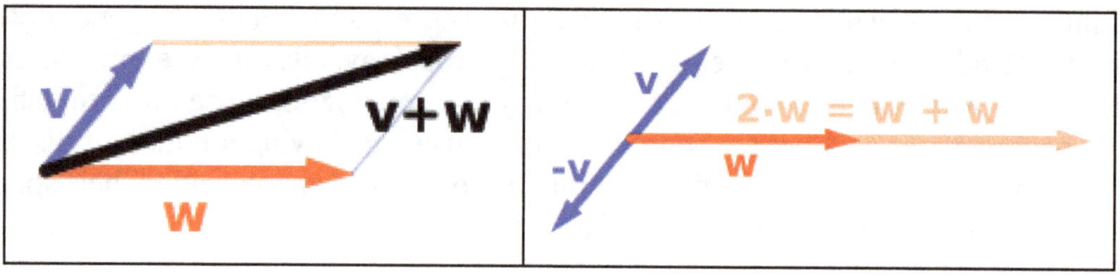

Second Example: Ordered Pairs of Numbers

A second key example of a vector space is provided by pairs of real numbers x and y. (The order of the components x and y is significant, so such a pair is also called an ordered pair.) Such a pair is written as (x, y). The sum of two such pairs and multiplication of a pair with a number is defined as follows:

$$(x_1, y_1) + (x_2, y_2) = (x_1 + x_2, y_1 + y_2)$$

and

$$a(x, y) = (ax, ay).$$

The first example above reduces to this one if the arrows are represented by the pair of Cartesian coordinates of their end points.

Definition

A vector space over a field F is a set V together with two operations that satisfy the eight axioms. Elements of V are commonly called *vectors*. Elements of F are commonly called *scalars*. The first operation, called *vector addition* or simply *addition*, takes any two vectors v and w and assigns to them a third vector which is commonly written as v + w, and called the sum of these two vectors. The second operation, called *scalar multiplication* takes any scalar a and any vector v and gives another vector av.

In this article, vectors are represented in boldface to distinguish them from scalars. In the two examples above, the field is the field of the real numbers and the set of the vectors consists of the planar arrows with fixed starting point and of pairs of real numbers, respectively.

To qualify as a vector space, the set V and the operations of addition and multiplication must adhere to a number of requirements called axioms. In the list below, let u, v and w be arbitrary vectors in V, and a and b scalars in F.

Axiom	Meaning
Associativity of addition	u + (v + w) = (u + v) + w
Commutativity of addition	u + v = v + u

Identity element of addition	There exists an element 0 ∈ V, called the zero vector, such that v + 0 = v for all v ∈ V.
Inverse elements of addition	For every v ∈ V, there exists an element −v ∈ V, called the additive inverse of v, such that v + (−v) = 0.
Compatibility of scalar multiplication with field multiplication	a(bv) = (ab)v
Identity element of scalar multiplication	1v = v, where 1 denotes the multiplicative identity in F.
Distributivity of scalar multiplication with respect to vector addition	a(u + v) = au + av
Distributivity of scalar multiplication with respect to field addition	(a + b)v = av + bv

These axioms generalize properties of the vectors introduced in the above examples. Indeed, the result of addition of two ordered pairs (as in the second example above) does not depend on the order of the summands:

$$(x_v, y_v) + (x_w, y_w) = (x_w, y_w) + (x_v, y_v).$$

Likewise, in the geometric example of vectors as arrows, v + w = w + v since the parallelogram defining the sum of the vectors is independent of the order of the vectors. All other axioms can be checked in a similar manner in both examples. Thus, by disregarding the concrete nature of the particular type of vectors, the definition incorporates these two and many more examples in one notion of vector space.

Subtraction of two vectors and division by a (non-zero) scalar can be defined as

$$v - w = v + (-w),$$

$$v/a = (1/a)v.$$

When the scalar field *F* is the real numbers R, the vector space is called a *real vector space*. When the scalar field is the complex numbers, it is called a *complex vector space*. These two cases are the ones used most often in engineering. The general definition of a vector space allows scalars to be elements of any fixed field *F*. The notion is then known as an *F-vector spaces* or a *vector space over F*. A field is, essentially, a set of numbers possessing addition, subtraction, multiplication and division operations.For example, rational numbers also form a field.

In contrast to the intuition stemming from vectors in the plane and higher-dimensional cases, there is, in general vector spaces, no notion of nearness, angles or distances. To deal with such matters, particular types of vector spaces are introduced.

Alternative Formulations and Elementary Consequences

Vector addition and scalar multiplication are operations, satisfying the closure property: u + v and av are in V for all a in F, and u, v in V. Some older sources mention these properties as separate axioms.

In the parlance of abstract algebra, the first four axioms can be subsumed by requiring the set of vectors to be an abelian group under addition. The remaining axioms give this group an F-module structure. In other words, there is a ring homomorphism f from the field F into the endomorphism ring of the group of vectors. Then scalar multiplication av is defined as $(f(a))$(v).

There are a number of direct consequences of the vector space axioms. Some of them derive from elementary group theory, applied to the additive group of vectors: for example the zero vector 0 of V and the additive inverse −v of any vector v are unique. Other properties follow from the distributive law, for example av equals 0 if and only if a equals 0 or v equals 0.

History

Vector spaces stem from affine geometry via the introduction of coordinates in the plane or three-dimensional space. Around 1636, Descartes and Fermat founded analytic geometry by equating solutions to an equation of two variables with points on a plane curve. In 1804, to achieve geometric solutions without using coordinates, Bolzano introduced certain operations on points, lines and planes, which are predecessors of vectors. His work was then used in the conception of barycentric coordinates by Möbius in 1827. In 1828 C. V. Mourey suggested the existence of an algebra surpassing not only ordinary algebra but also two-dimensional algebra created by him searching a geometrical interpretation of complex numbers.

The definition of vectors was founded on Bellavitis' notion of the bipoint, an oriented segment of which one end is the origin and the other a target, then further elaborated with the presentation of complex numbers by Argand and Hamilton and the introduction of quaternions and biquaternions by the latter. They are elements in R^2, R^4, and R^8; their treatment as linear combinations can be traced back to Laguerre in 1867, who also defined systems of linear equations.

In 1857, Cayley introduced matrix notation, which allows for a harmonization and simplification of linear maps. Around the same time, Grassmann studied the barycentric calculus initiated by Möbius. He envisaged sets of abstract objects endowed with operations. In his work, the concepts of linear independence and dimension, as well as scalar products, are present. In fact, Grassmann's 1844 work exceeds the framework of vector spaces, since his consideration of multiplication led him to what are today called algebras. Peano was the first to give the modern definition of vector spaces and linear maps in 1888.

An important development of vector spaces is due to the construction of function spaces by Lebesgue. This was later formalized by Banach and Hilbert, around 1920. At that time, algebra and the new field of functional analysis began to interact, notably with key concepts such as spaces of p-integrable functions and Hilbert spaces.

Vector spaces, including infinite-dimensional ones, then became a firmly established notion, and many mathematical branches started making use of this concept.

Examples

Coordinate Spaces

The simplest example of a vector space over a field F is the field itself, equipped with its standard addition and multiplication. More generally, a vector space can be composed of n-tuples (sequences of length n) of elements of F, such as

$(a_1, a_2, ..., a_n)$, where each a_i is an element of F.

A vector space composed of all the n-tuples of a field F is known as a *coordinate space*, usually denoted F^n. The case $n = 1$ is the above-mentioned simplest example, in which the field F is also regarded as a vector space over itself. The case $F = R$ and $n = 2$ was discussed in the introduction above.

Complex Numbers and Other Field Extensions

The set of complex numbers C, i.e., numbers that can be written in the form $x + iy$ for real numbers x and y where i is the imaginary unit, form a vector space over the reals with the usual addition and multiplication: $(x + iy) + (a + ib) = (x + a) + i(y + b)$ and $c \cdot (x + iy) = (c \cdot x) + i(c \cdot y)$ for real numbers x, y, a, b and c. The various axioms of a vector space follow from the fact that the same rules hold for complex number arithmetic.

In fact, the example of complex numbers is essentially the same (i.e., it is *isomorphic*) to the vector space of ordered pairs of real numbers mentioned above: if we think of the complex number $x + i\,y$ as representing the ordered pair (x, y) in the complex plane then we see that the rules for sum and scalar product correspond exactly to those in the earlier example.

More generally, field extensions provide another class of examples of vector spaces, particularly in algebra and algebraic number theory: a field F containing a smaller field E is an E-vector space, by the given multiplication and addition operations of F. For example, the complex numbers are a vector space over R, and the field extension $Q(i\sqrt{5})$ is a vector space over Q.

Function Spaces

Functions from any fixed set Ω to a field F also form vector spaces, by performing addition and scalar multiplication pointwise. That is, the sum of two functions f and g is the function $(f + g)$ given by

$$(f + g)(w) = f(w) + g(w),$$

and similarly for multiplication. Such function spaces occur in many geometric situations, when Ω is the real line or an interval, or other subsets of R. Many notions in topology and analysis, such as continuity, integrability or differentiability are well-behaved with respect to linearity: sums and scalar multiples of functions possessing such a property still have that property. Therefore, the set of such functions are vector spaces. They are studied in greater detail using the methods of functional analysis. Algebraic constraints also yield vector spaces: the *vector space F[x]* is given by polynomial functions:

$f(x) = r_0 + r_1x + ... + r_{n-1}x^{n-1} + r_nx^n$, where the coefficients r_0, ..., r_n are in F.

Linear Equations

Systems of homogeneous linear equations are closely tied to vector spaces. For example, the solutions of

$$a + 3b + c = 0$$

$$4a + 2b + 2c = 0$$

are given by triples with arbitrary a, $b = a/2$, and $c = -5a/2$. They form a vector space: sums and scalar multiples of such triples still satisfy the same ratios of the three variables; thus they are solutions, too. Matrices can be used to condense multiple linear equations as above into one vector equation, namely

$$Ax = 0,$$

where $A = \begin{bmatrix} 1 & 3 & 1 \\ 4 & 2 & 2 \end{bmatrix}$ is the matrix containing the coefficients of the given equations, x is the vector (a, b, c), Ax denotes the matrix product, and $0 = (0, 0)$ is the zero vector. In a similar vein, the solutions of homogeneous *linear differential equations* form vector spaces. For example,

$$f''(x) + 2f'(x) + f(x) = 0$$

yields $f(x) = ae^{-x} + bxe^{-x}$, where a and b are arbitrary constants, and e^x is the natural exponential function.

Basis and Dimension

Bases allow one to represent vectors by a sequence of scalars called *coordinates* or *components*. A basis is a (finite or infinite) set $B = \{b_i\}_{i \in I}$ of vectors b_i, for convenience often indexed by some index set I, that spans the whole space and is linearly independent. "Spanning the whole space" means that any vector v can be expressed as a finite sum (called a *linear combination*) of the basis elements:

$$v = a_1 b_{i_1} + a_2 b_{i_2} + \cdots + a_n b_{i_n}, \qquad (1)$$

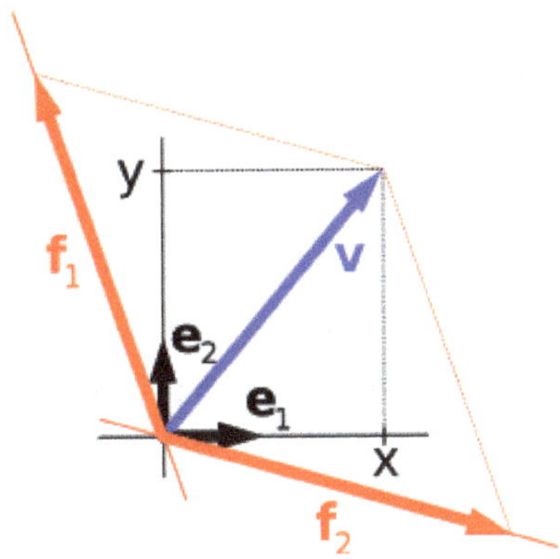

A vector v in R^2 (blue) expressed in terms of different bases: using the standard basis of R^2 $v = xe_1 + ye_2$ (black), and using a different, non-orthogonal basis: $v = f_1 + f_2$ (red).

where the a_k are scalars, called the coordinates (or the components) of the vector v with respect to the basis B, and b_{i_k} ($k = 1, ..., n$) elements of B. Linear independence means that the coordinates a_k are uniquely determined for any vector in the vector space.

For example, the coordinate vectors $e_1 = (1, 0, ..., 0)$, $e_2 = (0, 1, 0, ..., 0)$, to $e_n = (0, 0, ..., 0, 1)$, form a basis of F^n, called the standard basis, since any vector $(x_1, x_2, ..., x_n)$ can be uniquely expressed as a linear combination of these vectors:

$$(x_1, x_2, ..., x_n) = x_1(1, 0, ..., 0) + x_2(0, 1, 0, ..., 0) + ... + x_n(0, ..., 0, 1) = x_1e_1 + x_2e_2 + ... + x_ne_n.$$

The corresponding coordinates $x_1, x_2, ..., x_n$ are just the Cartesian coordinates of the vector.

Every vector space has a basis. This follows from Zorn's lemma, an equivalent formulation of the Axiom of Choice. Given the other axioms of Zermelo–Fraenkel set theory, the existence of bases is equivalent to the axiom of choice. The ultrafilter lemma, which is weaker than the axiom of choice, implies that all bases of a given vector space have the same number of elements, or cardinality (cf. *Dimension theorem for vector spaces*). It is called the *dimension* of the vector space, denoted dim V. If the space is spanned by finitely many vectors, the above statements can be proven without such fundamental input from set theory.

The dimension of the coordinate space F^n is n, by the basis exhibited above. The dimension of the polynomial ring $F[x]$ introduced above is countably infinite, a basis is given by $1, x, x^2, ...$ A fortiori, the dimension of more general function spaces, such as the space of functions on some (bounded or unbounded) interval, is infinite. Under suitable regularity assumptions on the coefficients involved, the dimension of the solution space of a homogeneous ordinary differential equation equals the degree of the equation. For example, the solution space for the above equation is generated by e^{-x} and xe^{-x}. These two functions are linearly independent over R, so the dimension of this space is two, as is the degree of the equation.

A field extension over the rationals Q can be thought of as a vector space over Q (by defining vector addition as field addition, defining scalar multiplication as field multiplication by elements of Q, and otherwise ignoring the field multiplication). The dimension (or degree) of the field extension Q(α) over Q depends on α. If α satisfies some polynomial equation

$$q_n\alpha^n + q_{n-1}\alpha^{n-1} + \ldots + q_0 = 0$$

with rational coefficients q_n, ..., q_0 (in other words, if α is algebraic), the dimension is finite. More precisely, it equals the degree of the minimal polynomial having α as a root. For example, the complex numbers C are a two-dimensional real vector space, generated by 1 and the imaginary unit *i*. The latter satisfies $i^2 + 1 = 0$, an equation of degree two. Thus, C is a two-dimensional R-vector space (and, as any field, one-dimensional as a vector space over itself, C). If α is not algebraic, the dimension of Q(α) over Q is infinite. For instance, for α = π there is no such equation, in other words π is transcendental.

Linear Maps and Matrices

The relation of two vector spaces can be expressed by *linear map* or *linear transformation*. They are functions that reflect the vector space structure—i.e., they preserve sums and scalar multiplication:

$f(x + y) = f(x) + f(y)$ and $f(a \cdot x) = a \cdot f(x)$ for all x and y in *V*, all *a* in *F*.

An *isomorphism* is a linear map $f : V \to W$ such that there exists an inverse map $g : W \to V$, which is a map such that the two possible compositions $f \circ g : W \to W$ and $g \circ f : V \to V$ are identity maps. Equivalently, *f* is both one-to-one (injective) and onto (surjective). If there exists an isomorphism between *V* and *W*, the two spaces are said to be *isomorphic*; they are then essentially identical as vector spaces, since all identities holding in *V* are, via *f*, transported to similar ones in *W*, and vice versa via *g*.

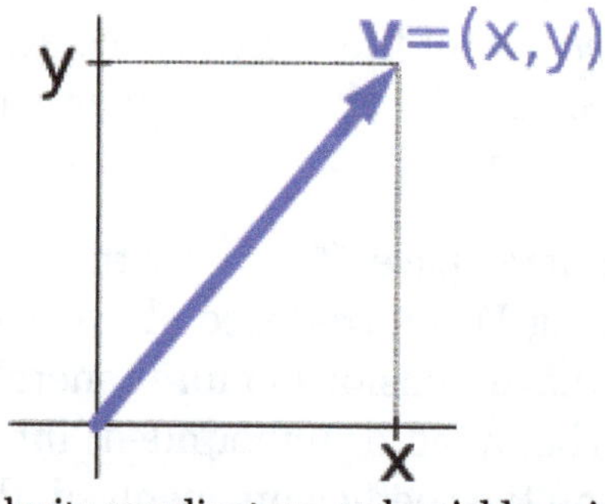

Describing an arrow vector **v** by its coordinates *x* and *y* yields an isomorphism of vector spaces.

For example, the “arrows in the plane” and “ordered pairs of numbers” vector spaces in the introduction are isomorphic: a planar arrow v departing at the origin of some (fixed) coordinate system can be expressed as an ordered pair by considering the *x*-

and y-component of the arrow, as shown in the image. Conversely, given a pair (x, y), the arrow going by x to the right (or to the left, if x is negative), and y up (down, if y is negative) turns back the arrow v.

Linear maps $V \to W$ between two vector spaces form a vector space $\mathrm{Hom}_F(V, W)$, also denoted $L(V, W)$. The space of linear maps from V to F is called the *dual vector space*, denoted V^*. Via the injective natural map $V \to V^{**}$, any vector space can be embedded into its *bidual*; the map is an isomorphism if and only if the space is finite-dimensional.

Once a basis of V is chosen, linear maps $f: V \to W$ are completely determined by specifying the images of the basis vectors, because any element of V is expressed uniquely as a linear combination of them. If $\dim V = \dim W$, a 1-to-1 correspondence between fixed bases of V and W gives rise to a linear map that maps any basis element of V to the corresponding basis element of W. It is an isomorphism, by its very definition. Therefore, two vector spaces are isomorphic if their dimensions agree and vice versa. Another way to express this is that any vector space is *completely classified* (up to isomorphism) by its dimension, a single number. In particular, any n-dimensional F-vector space V is isomorphic to F^n. There is, however, no "canonical" or preferred isomorphism; actually an isomorphism $\varphi : F^n \to V$ is equivalent to the choice of a basis of V, by mapping the standard basis of F^n to V, via φ. The freedom of choosing a convenient basis is particularly useful in the infinite-dimensional context.

Matrices

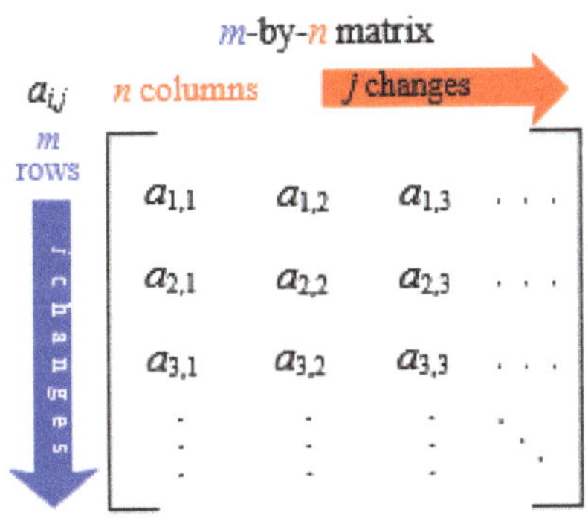

A typical matrix

Matrices are a useful notion to encode linear maps. They are written as a rectangular array of scalars as in the image at the right. Any m-by-n matrix A gives rise to a linear map from F^n to F^m, by the following

$$\mathrm{x} = (x_1, x_2, \cdots, x_n) \mapsto \left(\sum_{j=1}^{n} a_{1j}x_j, \sum_{j=1}^{n} a_{2j}x_j, \cdots, \sum_{j=1}^{n} a_{mj}x_j \right);$$ where denotes summation,

or, using the matrix multiplication of the matrix A with the coordinate vector x:

$$\boldsymbol{x} \mapsto A\boldsymbol{x}.$$

Moreover, after choosing bases of V and W, *any* linear map $f : V \to W$ is uniquely represented by a matrix via this assignment.

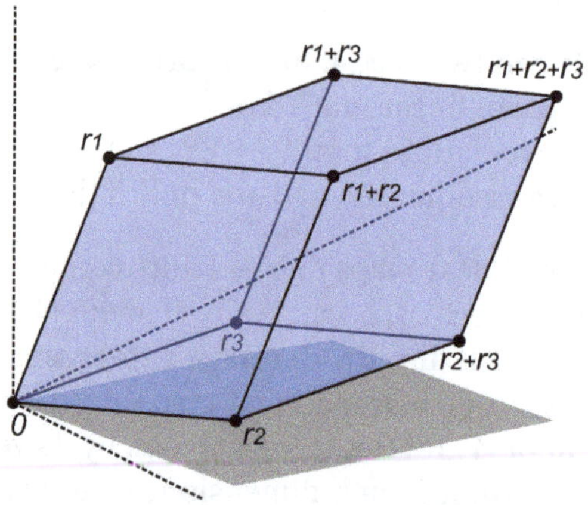

The volume of this parallelepiped is the absolute value of the determinant of the 3-by-3 matrix formed by the vectors r_1, r_2, and r_3.

The determinant det (A) of a square matrix A is a scalar that tells whether the associated map is an isomorphism or not: to be so it is sufficient and necessary that the determinant is nonzero. The linear transformation of R^n corresponding to a real n-by-n matrix is orientation preserving if and only if its determinant is positive.

Eigenvalues and Eigenvectors

Endomorphisms, linear maps $f : V \to V$, are particularly important since in this case vectors v can be compared with their image under f, f(v). Any nonzero vector v satisfying $\lambda v = f(v)$, where λ is a scalar, is called an eigenvector of f with eigenvalue λ. Equivalently, v is an element of the kernel of the difference $f - \lambda \cdot Id$ (where Id is the identity map $V \to V$). If V is finite-dimensional, this can be rephrased using determinants: f having eigenvalue λ is equivalent to

$$\det(f - \lambda \cdot \mathrm{Id}) = 0.$$

By spelling out the definition of the determinant, the expression on the left hand side can be seen to be a polynomial function in λ, called the characteristic polynomial of f. If the field F is large enough to contain a zero of this polynomial (which automatically happens for F algebraically closed, such as F = C) any linear map has at least one eigenvector. The vector space V may or may not possess an eigenbasis, a basis consisting of eigenvectors. This phenomenon is governed by the Jordan canonical form of the map.The set of all eigenvectors corresponding to a particular eigenvalue of f forms a vector space known as the *eigenspace* corresponding to the eigenvalue (and f) in question. To achieve the spectral theorem, the corresponding statement in the infinite-dimensional case, the machinery of functional analysis is needed.

Basic Constructions

In addition to the above concrete examples, there are a number of standard linear algebraic constructions that yield vector spaces related to given ones. In addition to the definitions given below, they are also characterized by universal properties, which determine an object X by specifying the linear maps from X to any other vector space.

Subspaces and Quotient Spaces

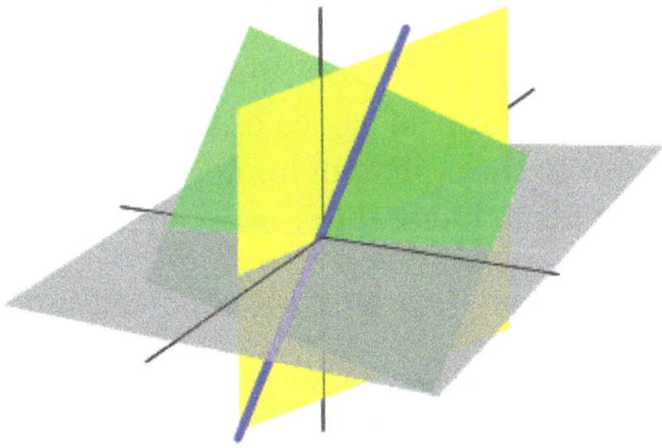

A line passing through the origin (blue, thick) in R^3 is a linear subspace. It is the intersection of two planes (green and yellow).

A nonempty subset W of a vector space V that is closed under addition and scalar multiplication (and therefore contains the 0-vector of V) is called a *linear subspace* of V, or simply a *subspace* of V, when the ambient space is unambiguously a vector space. Subspaces of V are vector spaces (over the same field) in their own right. The intersection of all subspaces containing a given set S of vectors is called its span, and it is the smallest subspace of V containing the set S. Expressed in terms of elements, the span is the subspace consisting of all the linear combinations of elements of S.

A linear subspace of dimension 1 is a vector line. A linear subspace of dimension 2 is a vector plane. A linear subspace that contains all elements but one of a basis of the ambient space is a vector hyperplane. In a vector space of finite dimension n, a vector hyperplane is thus a subspace of dimension $n - 1$.

The counterpart to subspaces are *quotient vector spaces*. Given any subspace $W \subset V$, the quotient space V/W ("V modulo W") is defined as follows: as a set, it consists of $\mathrm{v} + W = \{\mathrm{v} + \mathrm{w} : \mathrm{w} \in W\}$, where v is an arbitrary vector in V. The sum of two such elements $\mathrm{v}_1 + W$ and $\mathrm{v}_2 + W$ is $(\mathrm{v}_1 + \mathrm{v}_2) + W$, and scalar multiplication is given by $a \cdot (\mathrm{v} + W) = (a \cdot \mathrm{v}) + W$. The key point in this definition is that $\mathrm{v}_1 + W = \mathrm{v}_2 + W$ if and only if the difference of v_1 and v_2 lies in W. This way, the quotient space "forgets" information that is contained in the subspace W.

The kernel $\ker(f)$ of a linear map $f: V \to W$ consists of vectors v that are mapped to 0 in W. Both kernel and image $\mathrm{im}(f) = \{f(\mathrm{v}) : \mathrm{v} \in V\}$ are subspaces of V and W, respectively.

The existence of kernels and images is part of the statement that the category of vector spaces (over a fixed field F) is an abelian category, i.e. a corpus of mathematical objects and structure-preserving maps between them (a category) that behaves much like the category of abelian groups. Because of this, many statements such as the first isomorphism theorem (also called rank–nullity theorem in matrix-related terms)

$V / \ker(f) \equiv \operatorname{im}(f)$.

and the second and third isomorphism theorem can be formulated and proven in a way very similar to the corresponding statements for groups.

An important example is the kernel of a linear map $x \mapsto Ax$ for some fixed matrix A, as above. The kernel of this map is the subspace of vectors x such that $Ax = 0$, which is precisely the set of solutions to the system of homogeneous linear equations belonging to A. This concept also extends to linear differential equations

$a_0 f + a_1 \frac{df}{dx} + a_2 \frac{d^2 f}{dx^2} + \cdots + a_n \frac{d^n f}{dx^n} = 0$, where the coefficients a_i are functions in x, too.

In the corresponding map

$$f \mapsto D(f) = \sum_{i=0}^{n} a_i \frac{d^i f}{dx^i},$$

the derivatives of the function f appear linearly (as opposed to $f'(x)^2$, for example). Since differentiation is a linear procedure (i.e., $(f + g)' = f' + g'$ and $(c \cdot f)' = c \cdot f'$ for a constant c) this assignment is linear, called a linear differential operator. In particular, the solutions to the differential equation $D(f) = 0$ form a vector space (over R or C).

Direct Product and Direct Sum

The *direct product* of vector spaces and the *direct sum* of vector spaces are two ways of combining an indexed family of vector spaces into a new vector space.

The *direct product* $\prod_{i \in I} V_i$ of a family of vector spaces V_i consists of the set of all tuples $(v_i)_{i \in I}$, which specify for each index i in some index set I an element $\mathbf{v}_i$ of V_i. Addition and scalar multiplication is performed componentwise. A variant of this construction is the *direct sum* $\oplus_{i \in I} V_i$ (also called coproduct and denoted $\coprod_{i \in I} V_i$), where only tuples with finitely many nonzero vectors are allowed. If the index set I is finite, the two constructions agree, but in general they are different.

Tensor Product

The *tensor product* $V \otimes_F W$, or simply $V \otimes W$, of two vector spaces V and W is one of the central notions of multilinear algebra which deals with extending notions such as linear maps to several variables. A map $g : V \times W \to X$ is called bilinear if g is linear in both variables v and w. That is to say, for fixed w the map $v \mapsto g(v, w)$ is linear in the sense above and likewise for fixed v.

The tensor product is a particular vector space that is a *universal* recipient of bilinear maps g, as follows. It is defined as the vector space consisting of finite (formal) sums of symbols called tensors

$$v_1 \otimes w_1 + v_2 \otimes w_2 + \ldots + v_n \otimes w_n,$$

subject to the rules

$a \cdot (v \otimes w) = (a \cdot v) \otimes w = v \otimes (a \cdot w)$, *where a is a scalar,*

$(v_1 + v_2) \otimes w = v_1 \otimes w + v_2 \otimes w$, *and*

$v \otimes (w_1 + w_2) = v \otimes w_1 + v \otimes w_2$.

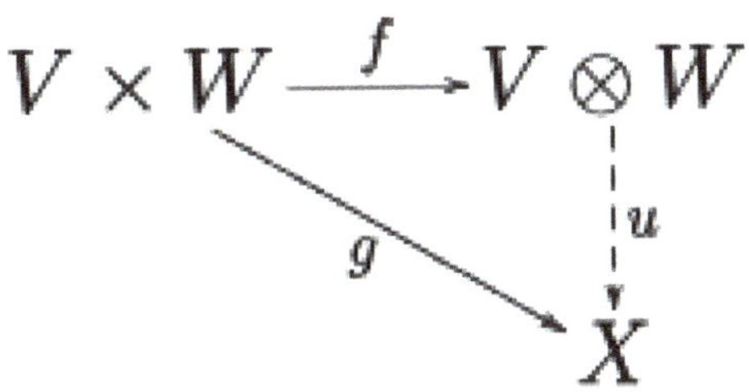

Commutative diagram depicting the universal property of the tensor product.

These rules ensure that the map f from the $V \times W$ to $V \otimes W$ that maps a tuple (v, w) to $v \otimes w$ is bilinear. The universality states that given *any* vector space X and *any* bilinear map $g : V \times W \to X$, there exists a unique map u, shown in the diagram with a dotted arrow, whose composition with f equals g: $u(v \otimes w) = g(v, w)$. This is called the universal property of the tensor product, an instance of the method—much used in advanced abstract algebra—to indirectly define objects by specifying maps from or to this object.

Vector Spaces with Additional Structure

From the point of view of linear algebra, vector spaces are completely understood insofar as any vector space is characterized, up to isomorphism, by its dimension. However, vector spaces *per se* do not offer a framework to deal with the question—crucial to analysis—whether a sequence of functions converges to another function. Likewise, linear algebra is not adapted to deal with infinite series, since the addition operation allows only finitely many terms to be added. *Therefore, the needs of functional analysis require considering additional structures.*

A vector space may be given a partial order $\leq$, under which some vectors can be compared. For example, n-dimensional real space R^n can be ordered by comparing its vectors componentwise. Ordered vector spaces, for example Riesz spaces, are fundamental to Lebesgue integration, which relies on the ability to express a function as a difference of two positive functions

$$f = f^+ - f^-,$$

where f^+ denotes the positive part of f and f^- the negative part.

Normed Vector Spaces and Inner Product Spaces

"Measuring" vectors is done by specifying a norm, a datum which measures lengths of vectors, or by an inner product, which measures angles between vectors. Norms and inner products are denoted $|\mathrm{v}|$ and $\langle \mathrm{v}, \mathrm{w} \rangle$, respectively. The datum of an inner product entails that lengths of vectors can be defined too, by defining the associated norm $|\mathrm{v}| := \sqrt{\langle \mathrm{v}, \mathrm{v} \rangle}$. Vector spaces endowed with such data are known as *normed vector spaces* and *inner product spaces*, respectively.

Coordinate space F^n can be equipped with the standard dot product:

$$\langle \mathrm{x}, \mathrm{y} \rangle = \mathrm{x} \cdot \mathrm{y} = \mathrm{x}_1 \mathrm{y}_1 + \cdots + \mathrm{x}_n \mathrm{y}_n.$$

In R^2, this reflects the common notion of the angle between two vectors x and y, by the law of cosines:

$$\mathrm{x} \cdot \mathrm{y} = \cos\left(\angle(\mathrm{x}, \mathrm{y})\right) \cdot |\mathrm{x}| \cdot |\mathrm{y}|.$$

Because of this, two vectors satisfying $\langle \mathrm{x}, \mathrm{y} \rangle = 0$ are called orthogonal. An important variant of the standard dot product is used in Minkowski space: R^4 endowed with the Lorentz product

$$\langle \mathrm{x} | \mathrm{y} \rangle = \mathrm{x}_1 \mathrm{y}_1 + \mathrm{x}_2 \mathrm{y}_2 + \mathrm{x}_3 \mathrm{y}_3 - \mathrm{x}_4 \mathrm{y}_4.$$

In contrast to the standard dot product, it is not positive definite: $\langle \mathrm{x} | \mathrm{x} \rangle$ also takes negative values, for example for $\mathrm{x} = (0, 0, 0, 1)$. Singling out the fourth coordinate—corresponding to time, as opposed to three space-dimensions—makes it useful for the mathematical treatment of special relativity.

Topological Vector Spaces

Convergence questions are treated by considering vector spaces *V* carrying a compatible topology, a structure that allows one to talk about elements being close to each other. Compatible here means that addition and scalar multiplication have to be continuous maps. Roughly, if x and y in *V*, and *a* in *F* vary by a bounded amount, then so do x + y and ax. To make sense of specifying the amount a scalar changes, the field *F* also has to carry a topology in this context; a common choice are the reals or the complex numbers.

In such *topological vector spaces* one can consider series of vectors. The infinite sum

$$\sum_{i=0}^{\infty} f_i$$

denotes the limit of the corresponding finite partial sums of the sequence $(f_i)_{i\in N}$ of elements of *V*. For example, the f_i could be (real or complex) functions belonging to some function space *V*, in which case the series is a function series. The mode of convergence of the series depends on the topology imposed on the function space. In such cases, pointwise convergence and uniform convergence are two prominent examples.

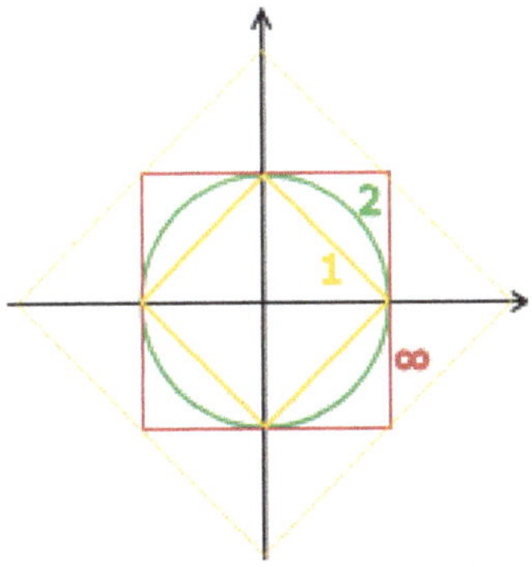

Unit "spheres" in R^2 consist of plane vectors of norm 1. Depicted are the unit spheres in different *p*-norms, for *p* = 1, 2, and ∞. The bigger diamond depicts points of 1-norm equal to 2.

A way to ensure the existence of limits of certain infinite series is to restrict attention to spaces where any Cauchy sequence has a limit; such a vector space is called complete. Roughly, a vector space is complete provided that it contains all necessary limits. For example, the vector space of polynomials on the unit interval [0,1], equipped with the topology of uniform convergence is not complete because any continuous function on [0,1] can be uniformly approximated by a sequence of polynomials, by the Weierstrass approximation theorem. In contrast, the space of *all* continuous functions on [0,1] with the same topology is complete. A norm gives rise to a topology by defining that a sequence of vectors v_n converges to v if and only if

$$\lim_{n\to\infty} | v_n - v |= 0.$$

Banach and Hilbert spaces are complete topological vector spaces whose topologies are given, respectively, by a norm and an inner product. Their study—a key piece of functional analysis—focusses on infinite-dimensional vector spaces, since all norms on finite-dimensional topological vector spaces give rise to the same notion of convergence. The image at the right shows the equivalence of the 1-norm and ∞-norm on R^2: as the unit "balls" enclose each other, a sequence converges to zero in one norm if and only if it so does in the other norm. In the infinite-dimensional case, however, there will generally be inequivalent topologies, which makes the study of topological vector spaces richer than that of vector spaces without additional data.

From a conceptual point of view, all notions related to topological vector spaces should match the topology. For example, instead of considering all linear maps (also called functionals) $V \to W$, maps between topological vector spaces are required to be continuous. In particular, the *(topological) dual space* V^* consists of continuous functionals $V \to \mathbb{R}$ (or to $\mathbb{C}$). The fundamental Hahn–Banach theorem is concerned with separating subspaces of appropriate topological vector spaces by continuous functionals.

Banach Spaces

Banach spaces, introduced by Stefan Banach, are complete normed vector spaces. A first example is the vector space ℓ^p consisting of infinite vectors with real entries $x = (x_1, x_2, \ldots)$ whose p-norm ($1 \le p \le \infty$) given by

$$|x|_p := \left(\sum_i |x_i|^p\right)^{1/p} \quad \text{for } p < \infty \text{ and } |x|_\infty := \sup_i |x_i|$$

is finite. The topologies on the infinite-dimensional space ℓ^p are inequivalent for different p. E.g. the sequence of vectors $x_n = (2^{-n}, 2^{-n}, \ldots, 2^{-n}, 0, 0, \ldots)$, i.e. the first 2^n components are 2^{-n}, the following ones are 0, converges to the zero vector for $p = \infty$, but does not for $p = 1$:

$$|x_n|_\infty = \sup(2^{-n}, 0) = 2^{-n} \to 0, \text{ but } |x_n|_1 = \sum_{i=1}^{2^n} 2^{-n} = 2^n \cdot 2^{-n} = 1.$$

More generally than sequences of real numbers, functions $f\colon \Omega \to \mathbb{R}$ are endowed with a norm that replaces the above sum by the Lebesgue integral

$$|f|_p := \left(\int_\Omega |f(x)|^p \, dx\right)^{1/p}.$$

The space of integrable functions on a given domain Ω (for example an interval) satisfying $|f|_p < \infty$, and equipped with this norm are called Lebesgue spaces, denoted $L^p(\Omega)$. These spaces are complete. (If one uses the Riemann integral instead, the space is *not* complete, which may be seen as a justification for Lebesgue's integration theory.) Concretely this means that for any sequence of Lebesgue-integrable functions $f_1, f_2, \ldots$ with $|f_n|_p < \infty$, satisfying the condition

$$\lim_{k,n\to\infty} \int_\Omega |f_k(x) - f_n(x)|^p \, dx = 0$$

there exists a function $f(x)$ belonging to the vector space $L^p(\Omega)$ such that

$$\lim_{k\to\infty} \int_\Omega |f(x) - f_k(x)|^p \, dx = 0.$$

Imposing boundedness conditions not only on the function, but also on its derivatives leads to Sobolev spaces.

Hilbert Spaces

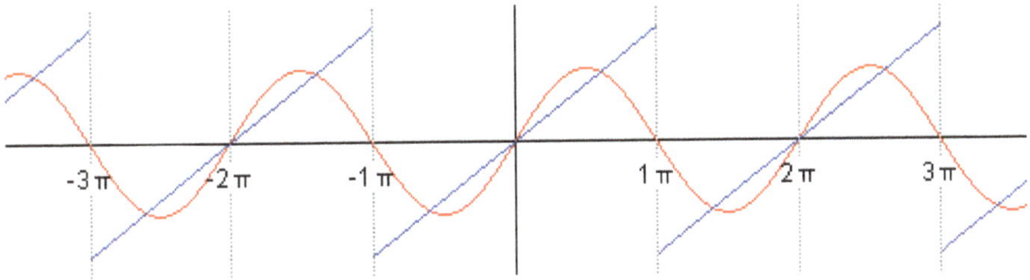

The succeeding snapshots show summation of 1 to 5 terms in approximating a periodic function (blue) by finite sum of sine functions (red).

Complete inner product spaces are known as *Hilbert spaces*, in honor of David Hilbert. The Hilbert space $L^2(\Omega)$, with inner product given by

$$\langle f, g \rangle = \int_{\Omega} f(x)\overline{g(x)}dx,$$

where $\overline{g(x)}$ denotes the complex conjugate of $g(x)$, is a key case.

By definition, in a Hilbert space any Cauchy sequence converges to a limit. Conversely, finding a sequence of functions f_n with desirable properties that approximates a given limit function, is equally crucial. Early analysis, in the guise of the Taylor approximation, established an approximation of differentiable functions f by polynomials. By the *Stone–Weierstrass theorem*, every continuous function on $[a, b]$ can be approximated as closely as desired by a polynomial. A similar approximation technique by trigonometric functions is commonly called Fourier expansion, and is much applied in engineering. *More generally, and more conceptually, the theorem yields a simple description of what "basic functions", or, in abstract Hilbert spaces, what basic vectors suffice to generate a Hilbert space H, in the sense that the closure of their span (i.e., finite linear combinations and limits of those) is the whole space.* Such a set of functions is called a *basis* of H, its cardinality is known as the Hilbert space dimension. Not only does the theorem exhibit suitable basis functions as sufficient for approximation purposes, but together with the Gram–Schmidt process, it enables one to construct a basis of orthogonal vectors. Such orthogonal bases are the Hilbert space generalization of the coordinate axes in finite-dimensional Euclidean space.

The solutions to various differential equations can be interpreted in terms of Hilbert spaces. For example, a great many fields in physics and engineering lead to such equations and frequently solutions with particular physical properties are used as basis functions, often orthogonal. As an example from physics, the time-dependent Schrödinger equation in quantum mechanics describes the change of physical properties in time by means of a partial differential equation, whose solutions are called wavefunctions. Definite values for physical properties such as energy, or momentum, correspond to eigenvalues of a certain (linear) differential operator and the associated wavefunctions are called eigenstates. The *spectral theorem decomposes a linear compact operator acting on functions in terms of these eigenfunctions and their eigenvalues.*

Algebras Over Fields

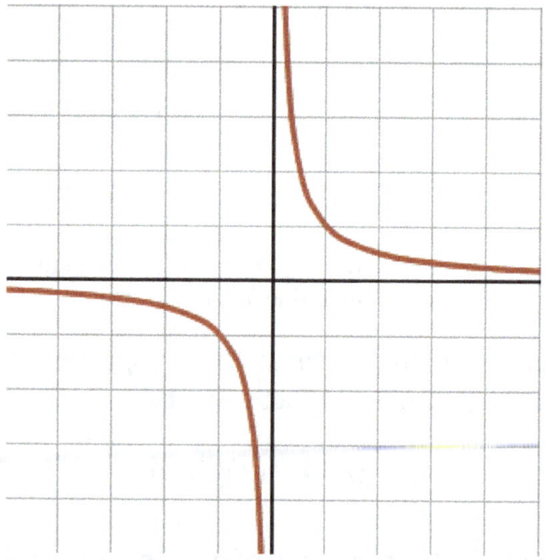

A hyperbola, given by the equation $x \cdot y = 1$. The coordinate ring of functions on this hyperbola is given by $R[x, y] / (x \cdot y - 1)$, an infinite-dimensional vector space over R.

General vector spaces do not possess a multiplication between vectors. A vector space equipped with an additional bilinear operator defining the multiplication of two vectors is an *algebra over a field*. Many algebras stem from functions on some geometrical object: since functions with values in a given field can be multiplied pointwise, these entities form algebras. The Stone–Weierstrass theorem mentioned above, for example, relies on Banach algebras which are both Banach spaces and algebras.

Commutative algebra makes great use of rings of polynomials in one or several variables, introduced above. Their multiplication is both commutative and associative. These rings and their quotients form the basis of algebraic geometry, because they are rings of functions of algebraic geometric objects.

Another crucial example are *Lie algebras*, which are neither commutative nor associative, but the failure to be so is limited by the constraints ($[x, y]$ denotes the product of x and y):

- $[x, y] = -[y, x]$ (anticommutativity), and
- $[x, [y, z]] + [y, [z, x]] + [z, [x, y]] = 0$ (Jacobi identity).

Examples include the vector space of n-by-n matrices, with $[x, y] = xy - yx$, the commutator of two matrices, and R^3, endowed with the cross product.

The tensor algebra T(V) is a formal way of adding products to any vector space V to obtain an algebra. As a vector space, it is spanned by symbols, called simple tensors

$\mathbf{v}_1 \otimes \mathbf{v}_2 \otimes \dots \otimes \mathbf{v}_n$, where the degree n varies.

The multiplication is given by concatenating such symbols, imposing the distributive law under addition, and requiring that scalar multiplication commute with the tensor product ⊗, much the same way as with the tensor product of two vector spaces introduced above. In general, there are no relations between $v_1 \otimes v_2$ and $v_2 \otimes v_1$. Forcing two such elements to be equal leads to the symmetric algebra, whereas forcing $v_1 \otimes v_2 = -v_2 \otimes v_1$ yields the exterior algebra.

When a field, *F* is explicitly stated, a common term used is *F*-algebra.

Applications

Vector spaces have many applications as they occur frequently in common circumstances, namely wherever functions with values in some field are involved. They provide a framework to deal with analytical and geometrical problems, or are used in the Fourier transform. This list is not exhaustive: many more applications exist, for example in optimization. The minimax theorem of game theory stating the existence of a unique payoff when all players play optimally can be formulated and proven using vector spaces methods. Representation theory fruitfully transfers the good understanding of linear algebra and vector spaces to other mathematical domains such as group theory.

Distributions

A *distribution* (or *generalized function*) is a linear map assigning a number to each "test" function, typically a smooth function with compact support, in a continuous way: in the above terminology the space of distributions is the (continuous) dual of the test function spacc. The latter space is endowed with a topology that takes into account not only *f* itself, but also all its higher derivatives. A standard example is the result of integrating a test function *f* over some domain Ω:

$$I(f) = \int_{\Omega} f(x)dx.$$

When $\Omega = \{p\}$, the set consisting of a single point, this reduces to the Dirac distribution, denoted by δ, which associates to a test function *f* its value at the *p*: $\delta(f) = f(p)$. Distributions are a powerful instrument to solve differential equations. Since all standard analytic notions such as derivatives are linear, they extend naturally to the space of distributions. Therefore, the equation in question can be transferred to a distribution space, which is bigger than the underlying function space, so that more flexible methods are available for solving the equation. For example, Green's functions and fundamental solutions are usually distributions rather than proper functions, and can then be used to find solutions of the equation with prescribed boundary conditions. The found solution can then in some cases be proven to be actually a true function, and a solution to the original equation (e.g., using the Lax–Milgram theorem, a consequence of the Riesz representation theorem).

Fourier Analysis

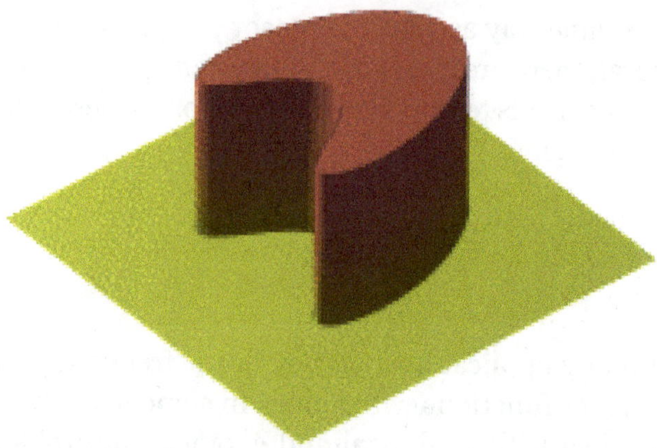

The heat equation describes the dissipation of physical properties over time, such as the decline of the temperature of a hot body placed in a colder environment (yellow depicts colder regions than red).

Resolving a periodic function into a sum of trigonometric functions forms a *Fourier series*, a technique much used in physics and engineering. The underlying vector space is usually the Hilbert space $L^2(0, 2\pi)$, for which the functions sin mx and cos mx (m an integer) form an orthogonal basis. The Fourier expansion of an L^2 function f is

$$\frac{a_0}{2}+\sum_{m=1}^{\infty}\left[a_m \cos(mx)+b_m \sin(mx)\right].$$

The coefficients a_m and b_m are called Fourier coefficients of f, and are calculated by the formulas

$$a_m=\frac{1}{\pi}\int_0^{2\pi} f(t)\cos(mt)dt \quad , \quad b_m=\frac{1}{\pi}\int_0^{2\pi} f(t)\sin(mt)dt.$$

In physical terms the function is represented as a superposition of sine waves and the coefficients give information about the function's frequency spectrum. A complex-number form of Fourier series is also commonly used. The concrete formulae above are consequences of a more general mathematical duality called Pontryagin duality. Applied to the group R, it yields the classical Fourier transform; an application in physics are reciprocal lattices, where the underlying group is a finite-dimensional real vector space endowed with the additional datum of a lattice encoding positions of atoms in crystals.

Fourier series are used to solve boundary value problems in partial differential equations. In 1822, Fourier first used this technique to solve the heat equation. A discrete version of the Fourier series can be used in sampling applications where the function value is known only at a finite number of equally spaced points. In this case the Fourier series is finite and its value is equal to the sampled values at all points. The set of coefficients is known as the discrete Fourier transform (DFT) of the given sample sequence. The DFT is one of the key tools of digital signal processing, a field whose applications include radar, speech encoding, image compression. The JPEG image format is an application of the closely related discrete cosine transform.

The fast Fourier transform is an algorithm for rapidly computing the discrete Fourier transform. It is used not only for calculating the Fourier coefficients but, using the convolution theorem, also for computing the convolution of two finite sequences. They in turn are applied in digital filters and as a rapid multiplication algorithm for polynomials and large integers (Schönhage-Strassen algorithm).

Differential Geometry

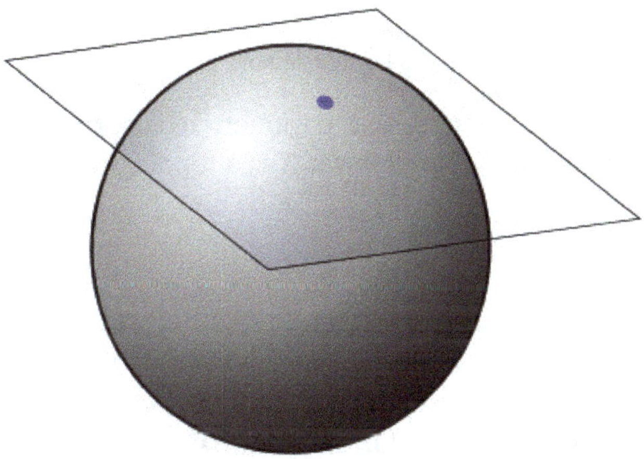

The tangent space to the 2-sphere at some point is the infinite plane touching the sphere in this point.

The tangent plane to a surface at a point is naturally a vector space whose origin is identified with the point of contact. The tangent plane is the best linear approximation, or linearization, of a surface at a point. Even in a three-dimensional Euclidean space, there is typically no natural way to prescribe a basis of the tangent plane, and so it is conceived of as an abstract vector space rather than a real coordinate space. The *tangent space* is the generalization to higher-dimensional differentiable manifolds.

Riemannian manifolds are manifolds whose tangent spaces are endowed with a suitable inner product. Derived therefrom, the Riemann curvature tensor encodes all curvatures of a manifold in one object, which finds applications in general relativity, for example, where the Einstein curvature tensor describes the matter and energy content of space-time. The tangent space of a Lie group can be given naturally the structure of a Lie algebra and can be used to classify compact Lie groups.

Generalizations

Vector Bundles

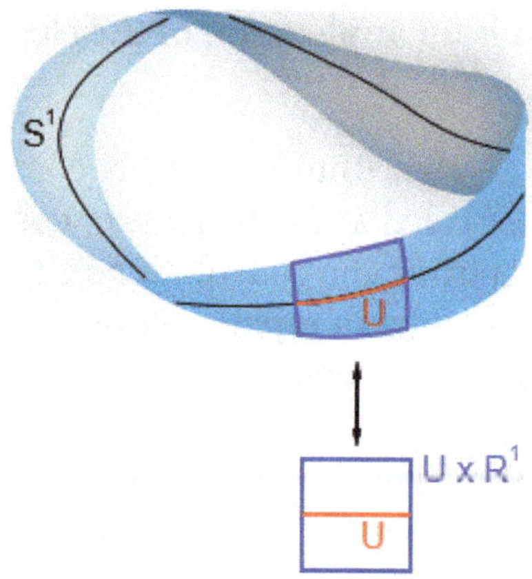

A Möbius strip. Locally, it looks like U × R.

A *vector bundle* is a family of vector spaces parametrized continuously by a topological space X. More precisely, a vector bundle over X is a topological space E equipped with a continuous map

$$\pi : E \to X$$

such that for every x in X, the fiber $\pi^{-1}(x)$ is a vector space. The case dim $V = 1$ is called a line bundle. For any vector space V, the projection $X \times V \to X$ makes the product $X \times V$ into a "trivial" vector bundle. Vector bundles over X are required to be locally a product of X and some (fixed) vector space V: for every x in X, there is a neighborhood U of x such that the restriction of π to $\pi^{-1}(U)$ is isomorphic to the trivial bundle $U \times V \to U$. Despite their locally trivial character, vector bundles may (depending on the shape of the underlying space X) be "twisted" in the large (i.e., the bundle need not be (globally isomorphic to) the trivial bundle $X \times V$). For example, the Möbius strip can be seen as a line bundle over the circle S^1 (by identifying open intervals with the real line). It is, however, different from the cylinder $S^1 \times$ R, because the latter is orientable whereas the former is not.

Properties of certain vector bundles provide information about the underlying topological space. For example, the tangent bundle consists of the collection of tangent spaces parametrized by the points of a differentiable manifold. The tangent bundle of the circle S^1 is globally isomorphic to $S^1 \times$ R, since there is a global nonzero vector field on S^1. In contrast, by the hairy ball theorem, there is no (tangent) vector field on the 2-sphere S^2 which is everywhere nonzero. K-theory studies the isomorphism classes of all vector bundles over some topological space. In addition to deepening topological and geometrical insight, it has purely algebraic consequences, such as the classification of finite-dimensional real division algebras: R, C, the quaternions H and the octonions O.

The cotangent bundle of a differentiable manifold consists, at every point of the manifold, of the dual of the tangent space, the cotangent space. Sections of that bundle are known as differential one-forms.

Modules

Modules are to rings what vector spaces are to fields: the same axioms, applied to a ring R instead of a field F, yield modules. The theory of modules, compared to that of vector spaces, is complicated by the presence of ring elements that do not have multiplicative inverses. For example, modules need not have bases, as the Z-module (i.e., abelian group) Z/2Z shows; those modules that do (including all vector spaces) are known as free modules. Nevertheless, a vector space can be compactly defined as a module over a ring which is a field with the elements being called vectors. Some authors use the term *vector space* to mean modules over a division ring. The algebro-geometric interpretation of commutative rings via their spectrum allows the development of concepts such as locally free modules, the algebraic counterpart to vector bundles.

Affine and Projective Spaces

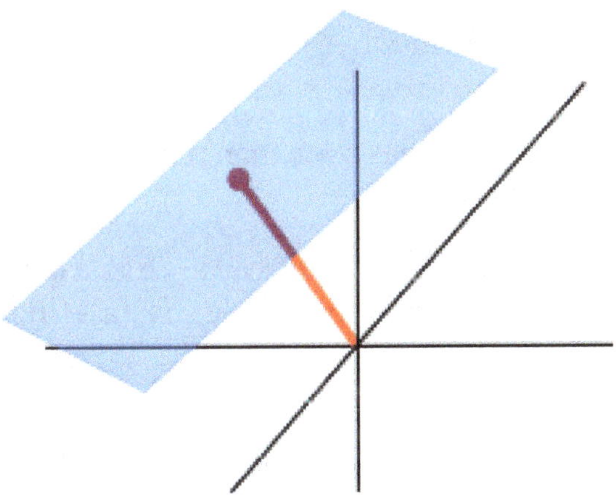

An affine plane (light blue) in R^3. It is a two-dimensional subspace shifted by a vector x (red).

Roughly, *affine spaces* are vector spaces whose origins are not specified. More precisely, an affine space is a set with a free transitive vector space action. In particular, a vector space is an affine space over itself, by the map

$$V \times V \to V, (v, a) \mapsto a + v.$$

If W is a vector space, then an affine subspace is a subset of W obtained by translating a linear subspace V by a fixed vector $x \in W$; this space is denoted by $x + V$ (it is a coset of V in W) and consists of all vectors of the form $x + v$ for $v \in V$. An important example is the space of solutions of a system of inhomogeneous linear equations

$$Ax = b$$

generalizing the homogeneous case b = 0 above. The space of solutions is the affine subspace x + V where x is a particular solution of the equation, and V is the space of solutions of the homogeneous equation (the nullspace of A).

The set of one-dimensional subspaces of a fixed finite-dimensional vector space V is known as *projective space*; it may be used to formalize the idea of parallel lines intersecting at infinity. Grassmannians and flag manifolds generalize this by parametrizing linear subspaces of fixed dimension k and flags of subspaces, respectively.

'Scalars' are going to be elements of a chosen (associative) ring with (identity) $\mathbb{K}$ and 'vectors' are going to be in general non commutative and will be equipped with a chosen involution but many significant results are best obtained in a setting when further restrictions are imposed on $\mathbb{K}$. We provide here a collection of definitions and examples, remarks, and other tit-bits, on our $\mathbb{K}$.

1. $\mathbb{Z}$, $\mathbb{Q}$, $\mathbb{R}$, $\mathbb{C}$ with the usual addition and multiplication are commutative rings. With the exception of $\mathbb{Z}$, they are fields,

2. Fix n > 0 in $\mathbb{Z}$ and write a ~ b iff n divides a – b, then

(i) [∵ n divides 0] a ~ a so this is reflexive ,

(ii) [∵ n divides a-b then n divides b–a] a ~ b ensures b ~ a so this is a symmetric, and

(iii) [∵ n divides a – b and n divides b – c then n divides a – c] a ~ b, b ~ c ensure a ~ c, so this is transitive.

Thus we have an equivalence relation on $\mathbb{Z}$; we read a ~ b as 'a is equivalent to b modulo n'. Write $[a] := \{b \in \mathbb{Z} \mid a \sim b\}$ and define $\mathbb{Z}_n := \{[a] \mid a \in \mathbb{Z}\}, [a]+[b] := [a+b], [a][b] := [ab]$.

(i) If $a' \in [a], b' \in [b]$ then n divides $a - a'$ as well as $b - b'$; therefore n divides $(a+b)-(a'-b')$. Thus $[a+a'] = [b+b']$ if $[a]=[a']$ and $[b]=[b']$. This means addition above is well defined. Similarly, if $a' \in [a]$, say $b = np + b'$, $b' \in [b]$, say $b = nq \; + b'$ then $ab = n^2pq + nqb' + nqa' + a'b'$ i.e. $a' - b' = n[npq + qb' + a'p]$ i.e. $a'b' \in [ab]$ then $[ab] = [a'b']$ if $[a]=[a']$ and $[b]=[b']$ which means multiplication is well defined.

(ii) Clearly $[a]+[b] = [a+b] = [b+a] = [b]+[a]$ and $0 = \{nq \mid q \in \mathbb{Z}\}$ so that if $a' \in [a]$ say $a = np + a'$ and $nq \in [0]$, we have $a' + nq = a - np + nq = a + n(q-p) \in [a]$ then $[a] + [0] = [a] = [0] + [a]$. Further $-a = -np - a'$ i.e. $[a] + [-a] = [0]$ which means $[-a] = -[a]$. Therefore, under addition as defined, $\mathbb{Z}_n$ an abelian group [it is clear that $[a] + ([b] + [c]) = [(a + b) + c] = ([a + b]) + [c] = ([a] + [b]) + [c]$].

(iii) If [a] = [1] then a = nq + 1 so that ba = bnq + b hence [ba] = [b] and ab = nqb + b hence [ab] = [b]. Thus [a] = [a] = [a] for each $[a] \in \mathbb{Z}_n$. Clearly [a]([b][c]) = [a][bc] = [a(bc)] = [(ab)c] = [ab][c] = ([a][b])[c] for all $[a],[b],[c] \in \mathbb{Z}_n$. Therefore under multiplication as defined, $\mathbb{Z}_n$ is a monoid with [1] as identity.

(iv) We have [a]([b] + [c]) = [a][b + c] = [a(b + c)] = [ab + ac] = [ab] + [ac] = [a][b] + [a][c] and ([a] + [b])[c] = [a + b][c] = [(a + b)c] = [ac + bc] = [ac] + [bc] = [a][c] + [b][c].Thus the distributive law hold.

(v) These calculations show that $\mathbb{Z}_n$ is a ring under the defined operations.

(vi) Since [a][b] = [ab] = [ba] for each [a], [b] ∈ $\mathbb{Z}_n$, we see that $\mathbb{Z}_n$ is a commutative ring.

(vii) [a][b] = 1 iff ab = nq + 1. Conversely, if there exist integers c, d with ac = nd + 1, let c = nq + r, $0 \le r < n$ so that nd + 1 = ac = anq + ar, i.e. ar = n(d – aq) + 1 and thus [a][r] = [ar] = [1]. Thus $[a] \in \mathbb{Z}_n$ is invertible iff a is relatively prime to n i.e. gcd(a, n) = 1, i.e. there exist integers c, d with ac – nd = 1; the multiplicative inverse of [a] is then given by [c] and we have $[c] = [a]^{-1}$ iff there exists an integer d such that ac – nd = 1

(viii) Therefore $\mathbb{Z}_n$ is a field iff n is prime.

3. The ring $\mathbb{Z}$ occupies a very special position: $\mathbb{Z} \to^{f} \mathbb{K}$ defined by f (n):= n1 = 1+. . .+1[1∈ $\mathbb{K}$] is the single-morphism for any ring $\mathbb{K}$ from $\mathbb{Z}$. [Indeed, $f(1) = 1_{\mathbb{K}}$ is must and so is $f(1) = f(1+...1) = f(1)+f(1)+...f(1) = 1_{\mathbb{K}}+...1_{\mathbb{K}} = n1_{\mathbb{K}}$ so $f(n) := n1_{\mathbb{K}}$ is the only possible formula if f is a ring-morphism and it is certainly clear that this formula supplies a ring-morphism $\mathbb{Z} \to \mathbb{K}$ for any ring $\mathbb{K}$; here we wrote the identity of $\mathbb{K}$ as $1_{\mathbb{K}}$ for extra clarity because $1 \in \mathbb{Z}$ is also there in the calculation but one usually writes $1_{\mathbb{K}}$ simply as 1].

We say "$\mathbb{Z}$ is an initial object in the category of rings" but this language will not be used in this course.

(i) The least integer k ≥ 0 such that f (k) = 0 is called the characteristic of $\mathbb{K}$; clearly $f = \{km \mid m \in \mathbb{Z}\}$.

(ii) (a) $\mathbb{Z}$, $\mathbb{Q}$, $\mathbb{R}$, $\mathbb{C}$ have characteristic 0.

(b) $\mathbb{Z}_n$ has characteristic n. [$\because [0] = \{nm \mid m \in \mathbb{Z}\}$]

(c) $\mathbb{Z}_{\ltimes} \times \mathbb{Z}_{>}$ has characteristic lcm(m, n).

(d) $\mathbb{K}$ has characteristic 1 iff f (1) = 0 i.e. 1=0 in $\mathbb{K}$. Since we have agreed that we have at least two distinct scalars $0 \neq 1$, no ring of scalars for this course will have characteristic 1 but it should be noted that not every textbook on linear algebra takes this position.

4. (i) In connection with 3(ii)(d) above, it is perhaps here we should note that some text- books on linear algebra accept rings without identity can be enlarged to a ring with identity so in the good old days one always did it. With modern development, there are situations [like 'K-theory'] in which rings or rather algebras, without identity play an essential role. But we accept rings only with identity.

(ii) On the other hand, many text books on linear algebra take only commutative rings as scalars and most of them in fact take only fields. Some accept only division rings. The basic trouble with these in as follows: Even if one works only with fields as scalars, one has to works with field as scalars, one has to work with 'block matrices' in which are non-commutative so linear algebra with non-commutative rings also comes into picture.

(iii) Working with only division rings or fields, one can avoid the term 'modula' and use only 'vector spaces'. This is an economy which is only apparent: practically all the work has to be done anyway. In particular, which one crucial fact: a linear operator $V \to^{T} V$ [where v is a vector space over field $\mathbb{F}$] is in fact just a modulo over the polynomial ring $\mathbb{F}$ [θ], is never mentioned, all the computational work has still to be done. So all one gains is that a concept has not been given its proper name.

(iv) We note [but will not prove] that (i) Every module over a division ring a free module (ii) If every module over $\mathbb{K}$-modulo, $\mathbb{K}$ must be a division ring. Thus the best result are certainly developed in the context of vector spaces [which are just modules over division rings] and if one uses the very convenient characterization in terms of determinant [of matrices] then it is almost essential to work with "finite-dimensional vector spaces over fields". However, it becomes an unnecessary journey into surprises later when one fields that many things do not work in more matrices are not invertible, that AB = 0 does not mean that A = 0 or B = 0, that AB = BA may not be true, and so on; but it does not mean that one should never move out of the world of real numbers.

(v) Some textbooks deal with only complex numbers as scalars; this is to prepare the student for Applicable functional analysis and hence to many rich and exciting subjects like quantum mechanics and Wavelets differential equations and so (classical) computation, one frequently meets with $\mathbb{Z}_n$ and since there is hardly any saving in learning efforts by declaring "by a scalar we mean a real number or a complex number" we have avoided this declaration.

(vi) Clearly, the 'choice of reasonable scalars' is a subjective choice. We hope our option of working with a non-commutative but associative ring with identity with identity equipped with a preferred conjugation prepares the users for a number of situations that may arise in subsequent mathematical education.

(vii) In all this, let us note that in $\mathbb{Z}$ we have [n] = [0] so that, for instance, $\lambda \in \mathbb{Z}_2$, $\mu \in \mathbb{Z}_2$ does not ensure that $\frac{\lambda+\mu}{2} \in \mathbb{Z}_2$. Many results will need "let F be a field of characteristic not equal to 2". [There are fields other than $\mathbb{Z}_2$ which have characteristic 2].

5. Let us note that the collection of all vectors in $\mathbb{R}^3$ is a ring with cross product as multi- plication since a × (b + c) = a × b + a × c. This not associative, [a × (b × c) = (a × b) × c fails] and dose not commutative, [a × b = b × a fails] and does not have an identity.

6. If $a = a_0 + a_1e_1 + a_2e_2 + a_3e_3$ is a quaternion then for any $r \in \mathbb{R}$ we clearly have ra = ar. Then $\mathbb{R} \subseteq \text{cen}\,\mathbb{H}$ [$\mathbb{H}$ the division ring of quaternions]. Conversely, suppose $q = (q_0 + q_1e_1 + q_2e_2 + q_3e_3) \in \text{cen}\ \mathbb{H}$ so that qa = aq for each $a \in \mathbb{H}$. Then we have $qe_1 = e_1q$ in particular. Now $qe_1 = q_0e_1 + q_1e_1^2 + q_2e_2e_1 + q_3e_3e_1 = -q_1 + q_0e_1 + q_3e_2 - q_2e_3$, while $e_1q =$

$e_1q_0 + e_1q_1e_1 + e_1q_2e_2 + e_1q_3e_3 = q_0e_1 + q_1e_1e_1 + q_2e_1e_2 + q_3e_1e_3$ [$q_i \in \mathbb{R} \subseteq \text{cen}\ \mathbb{H}$] $= -q_1 + q_0e_1 - q_3e_2 + q_2e_3$ which provides $q_3 = 0 = q_2$ $\left[\because \sum_{i=0}^{3} q_ie_i = \sum_{i=0}^{3} q_i'e_i \text{ iff } q_i = q_i' \text{ for each } i\right]$.

Thus $q = q_0 + q_1e_1$. We also have $qe_2 = e_2q$, i.e. $q_0e_2 + q_1e_1e_2 = e_2q_0 + e_2q_1e_1 = q_0e_2 + q_1e_2e_1$ i.e. $q_0e_2 + q_1e_3 = q_0e_2 - q_1e_3$ which means $q_1 = 0$. Thus we have $q = q_0 \in \mathbb{R}$ i.e. cen ($\mathbb{H}$) $\subseteq \mathbb{R}$. To sum up : cen $\mathbb{H} = \mathbb{R}$.

[We wrote a quaternion as $\sum_{i=0}^{3} q_ie_i$ rather than $\sum q_ie_i$ but this does not matter since $\mathbb{H}$ is an algebra over $\mathbb{R}$ and $\mathbb{R}$ is commutative].

7. Our rings are rings are rings with a given involutions. We took $\bar{\lambda} = \lambda_0 - e_1\lambda_1 - e_2\lambda_2 - e_3\lambda_3$ for $\lambda = \lambda_0 + e_1\lambda_1 + e_2\lambda_2 + e_3\lambda_3 \in \mathbb{H}$ as the standard involution but $\lambda^* = \lambda_0 + e_1\lambda_1 - e_2\lambda_2 - e_3\lambda_3$ is also an involution and there could be others. Let us look at some situations.

(i) (a) If $\mathbb{Z} \to^f \mathbb{Z}$ is a ring morphism, we have f (1) = 1 and thus f (m) = f (m.1) = mf (1) for all $m \in \mathbb{Z}$.

(b) If $\mathbb{Z}_n \to^f \mathbb{Z}_n$ is a ring morphism, we have $f[a] = f(a[1]) = af[1] = a[1] = [a]$..

(c) If $\mathbb{Q} \to^f \mathbb{Q}$ is a ring morphism, we have $nf\left(\frac{m}{n}\right) = f\left(n.\frac{m}{n}\right) = f(m) = f(m.1) = mf(1) = m = n.\frac{m}{n}$ and hence $f(\frac{m}{n}) = f(\frac{m}{n})$ for any $r \in \mathbb{Q}, r = \dfrac{m}{n}, m, n \in \mathbb{Z}, n > 0$.

(d) Suppose $\mathbb{K}$ and $\mathbb{L}$ are rings, $\mathbb{K} \to^f \mathbb{L}$ a ring morphism. Write ker f := {λ ∈ K/f (λ) = 0} Then if λ ∈ kerf and $a \in \mathbb{K}$, we have f (λa) = f (λ)f (a) = f (a) = 0 i.e. λa ∈ ker f ; similarly aλ ∈ kerf [This makes ker f an ideal of $\mathbb{K}$; for the definition of 'ideal] Now if $\mathbb{K}$ and $\mathbb{L}$ are division rings , $0 \neq \lambda \in \ker f$, then $1_K = \lambda^{-1}\lambda \in$ ker f so that $f(1_{\mathbb{K}}) = 1_{\mathbb{L}}$ and $1 \neq 0$ in any (division) ring. Therefore ker f = 0. Hence f is injective [f (λ) = f (μ) means f (λ − μ) = f (λ) − f (μ) = 0 which ensures λ − μ ∈ ker f i.e. λ − μ = 0 i.e. λ = μ].

To sum up: If $\mathbb{K}$ and $\mathbb{L}$ are division rings (and in particular, if they are fields), any ring morphism $\mathbb{K} \to^f \mathbb{L}$ is necessarily injective.

(ii) If $\mathbb{R} \to^{f} \mathbb{R}$ is a ring morphism, then by above it is injective and hence [by (i) c] we have f (r) = r for $r \in \mathbb{Q}$. If $x \in \mathbb{R}, x > 0$ then $f(x) = f\left(\sqrt{x}\sqrt{x}\right) = \left(f\left(\sqrt{x}\right)\right)^2$ so that if a > b and thus a – b > 0, we get f (a – b) = f (a) – f (b) > 0 i.e. f (a) > f (b). This means f must be order-preserving. Now let $x \in \mathbb{R}$ and choose r_n, $s_n \in \mathbb{Q}$ such that $r_n < x < s_n$ with $\cap_{n=1}^{\infty}[r_n, s_n] = \{x\}$

[This is possible because of the properties of $\mathbb{R}$]

Then we have $f(r_n) = r_n < f(x) < f(s_n) = s_n$ which means $f(x) \in \cap_{n=1}^{\infty}[r_n, s_n] = \{x\}$ and proves that f (x) = x.

Together with (i), (iii) shows that:

None of the rings $\mathbb{Z}$, $\mathbb{Z}_n$, $\mathbb{Q}$, $\mathbb{R}$ admits any ring morphism from itself to itself other than the identity. Since each of them is commutative, morphisms are the same as anti-morphisms. Therefore

No nontrivial involution, isomorphisms are the same as anti-isomorphism. If $\mathbb{C} \to^{f} \mathbb{C}$ is an isomorphism such that f (x) = x for $x \in \mathbb{R}$, then f (x + iy) = f (x) + f (i)f (y) with $(f(i))^2 = f(i)f(i) = f(i^2) = f(-1) = -1$ so that f (i) = ±i. To sum up: The only nontrivial involution on $\mathbb{C}$ with f (x) = x for $x \in \mathbb{R}$ is $x + iy \mapsto x - iy$.

(iii) If $d \neq 0.1$ is an integer which is square free [in the sense that its prime factorization has no square], we write $\mathbb{Z}\left[\sqrt{d}\right] := \left\{a + b\sqrt{d} \,|\, a, b \in \mathbb{Z}\right\}$ and $\mathbb{Q}\left[\sqrt{d}\right] := \left\{a + b\sqrt{d} \,|\, a, b \in \mathbb{Q}\right\}$.

Then under obvious multiplication and addition, there are commutative rings and $\mathbb{Q}\left[\sqrt{d}\right]$ is actually a field; both carry the non-trivial involution $a + b\sqrt{d} \mapsto a - b\sqrt{d}$.

(iv) For a prime number p, $\mathbb{Z}_p$ is a finite field with exactly p element; this has been introduced in 2(viii) above. We supply the following facts without proof; the proof are not exactly obvious and are properly speaking part of an algebra course which pays some attention to finite fields.

(a) A finite field has exactly p^r elements, p a prime, r > 0 an integer.

(b) There is exactly one finite field with p^r elements

(c) For the finite field with p^r elements, the characteristic is p. Further, $\lambda \mapsto \lambda^p$ is a ring morphism and $(\lambda + \mu)^p = \lambda^p + \mu^p$ holds.

(d) Thus on the field $\mathbb{F}$ with p^2 elements, $x \mapsto \overline{x} := x^p$ is an involution.

(e) Let us examine the field $\mathbb{F}$ with p^2 elements for p = 2. Quite clearly, this cannot be $\mathbb{Z}_4$ [in fact the field with p^r elements, r > 1, will never be $\mathbb{Z}_q$, $q = p^r$] because

$\mathbb{Z}_n$ is a field iff p is a prime. Verify that it carries addition and multiplication given by the tables

+	0	1	a	b
0	0	1	a	b
1	1	0	b	a
a	a	b	0	1
b	b	a	1	0

and

.	0	1	a	b
0	0	0	0	0
1	0	1	a	b
a	o	a	b	1
b	0	b	1	a

Its multiplicative fragment {1, a, b} must be an abelian group since it is a field.

Verify that this is same as the additive fragment of $\mathbb{Z}_3$. There are no finite division rings other than finite fields.

8. Consider the collection $\mathbb{K} := \{\begin{pmatrix} \alpha & \beta \\ -\bar{\beta} & \bar{\alpha} \end{pmatrix} | \alpha, \beta \in \mathbb{C}\}$

Since $\begin{pmatrix} \alpha & \beta \\ -\bar{\beta} & \bar{\alpha} \end{pmatrix} + \begin{pmatrix} \gamma & \delta \\ -\bar{\delta} & \bar{\gamma} \end{pmatrix} = \begin{pmatrix} \alpha+\gamma & \beta+\delta \\ -\bar{\beta}-\bar{\delta} & \bar{\alpha}+\bar{\gamma} \end{pmatrix} = \begin{pmatrix} \alpha+\gamma & \beta+\delta \\ -\overline{\beta+\delta} & \overline{\alpha+\gamma} \end{pmatrix}$,

$$\begin{pmatrix} \alpha & \beta \\ -\bar{\beta} & \bar{\alpha} \end{pmatrix}\begin{pmatrix} \gamma & \delta \\ -\bar{\delta} & \bar{\gamma} \end{pmatrix} = \begin{pmatrix} \alpha\gamma-\beta\bar{\delta} & \alpha\delta+\beta\bar{\gamma} \\ -\bar{\beta}\gamma-\bar{\alpha}\bar{\delta} & -\bar{\beta}\delta+\bar{\alpha}\,\bar{\gamma} \end{pmatrix} = \begin{pmatrix} \alpha\gamma-\beta\bar{\delta} & \alpha\delta+\beta\bar{\gamma} \\ -(\overline{\alpha\delta+\beta\bar{\gamma}}) & \overline{\alpha\gamma-\beta\bar{\delta}} \end{pmatrix},$$

We find that $\mathbb{K}$ is closed under the usual matrix addition and multiplication and thus obviously a ring [—the distributive law flows from the usual matrix situation $\text{Mat}_2(\mathbb{C})$] which is a sub ring of $\text{Mat}_2(\mathbb{C})$ because $\begin{pmatrix} 1 & 0 \\ 0 & 1 \end{pmatrix} = \text{Id}_2 \in \mathbb{K}$.

(i) We write $J := \begin{pmatrix} 0 & 1 \\ -1 & 0 \end{pmatrix}$, $K = \begin{pmatrix} 0 & i \\ i & 0 \end{pmatrix}, \begin{pmatrix} i & 0 \\ 0 & -i \end{pmatrix}$ and note

$$JK = \begin{pmatrix} 0 & 1 \\ -1 & 0 \end{pmatrix}\begin{pmatrix} 0 & i \\ i & 0 \end{pmatrix} = \begin{pmatrix} i & 0 \\ 0 & -i \end{pmatrix} = I,$$

$$KJ = \begin{pmatrix} 0 & i \\ i & 0 \end{pmatrix} \begin{pmatrix} 0 & 1 \\ -1 & 0 \end{pmatrix} = \begin{pmatrix} -i & 0 \\ 0 & i \end{pmatrix} - I,$$

$$KI = \begin{pmatrix} 0 & i \\ i & 0 \end{pmatrix} \begin{pmatrix} i & 0 \\ 0 & -i \end{pmatrix} = \begin{pmatrix} 0 & 1 \\ -1 & 0 \end{pmatrix} = J$$

$$IK = \begin{pmatrix} i & 0 \\ 0 & -i \end{pmatrix} \begin{pmatrix} 0 & i \\ i & 0 \end{pmatrix} = \begin{pmatrix} 0 & -1 \\ 1 & 0 \end{pmatrix} = -J$$

$$IJ = \begin{pmatrix} i & 0 \\ 0 & -i \end{pmatrix} \begin{pmatrix} 0 & 1 \\ -1 & 0 \end{pmatrix} = \begin{pmatrix} 0 & i \\ i & 0 \end{pmatrix} = K$$

$$JI = \begin{pmatrix} 0 & 1 \\ -1 & 0 \end{pmatrix} \begin{pmatrix} i & 0 \\ 0 & -i \end{pmatrix} = \begin{pmatrix} 0 & -i \\ -i & 0 \end{pmatrix} = -K$$

$$J^2 = \begin{pmatrix} 0 & 1 \\ -1 & 0 \end{pmatrix} \begin{pmatrix} 0 & 1 \\ -1 & 0 \end{pmatrix} = \begin{pmatrix} -1 & 0 \\ 0 & -1 \end{pmatrix} = -Id_2$$

$$K^2 = \begin{pmatrix} 0 & i \\ i & 0 \end{pmatrix} \begin{pmatrix} 0 & i \\ i & 0 \end{pmatrix} = \begin{pmatrix} -1 & 0 \\ 0 & -1 \end{pmatrix} = -ID_2$$

$$I^2 = \begin{pmatrix} i & 0 \\ 0 & -i \end{pmatrix} \begin{pmatrix} i & 0 \\ 0 & -i \end{pmatrix} = \begin{pmatrix} -1 & 0 \\ 0 & -1 \end{pmatrix} = -Id_2$$

If $\alpha = a + bi$, $\beta = c + di$, $a, b, c, d \in \mathbb{R}$ we can write a quaternion uniquely as $q = aId_2 + bI + cJ + dK$ and conversely a matrix of the sort aId2 + bI + cJ + dK would stand for the quaternion This can be done by correspondence $e_0 \leftrightarrow Id_2$, $e_1 \leftrightarrow I$, $e_2 \leftrightarrow J$, $e_3 \leftrightarrow K$ The calculation above show that the multiplication table for quaternions corresponds to the matrix multiplication performed above; since the addition obviously dose the mapping $f(e_0) = 1d_2$, $f(e_1) = I$, $f(e_2) = J$, $f(e_3) = K$ establishes an isomorphism of the division

algebra of the quaternions onto the matrix algebra introduced just now, so the two are the same. Further $f(\alpha) = \begin{pmatrix} \alpha & 0 \\ 0 & \bar{\alpha} \end{pmatrix} = aId_2 + bI$ for $\alpha = a + bi \in \mathbb{C}$ embeds the field $\mathbb{C}$ into $\mathbb{H}$ via $a \mapsto a + bi = \alpha = \begin{pmatrix} \alpha & 0 \\ 0 & \bar{\alpha} \end{pmatrix}$.

Similarly, with $a \in \mathbb{R}$ being identified to $\begin{pmatrix} a & 0 \\ 0 & a \end{pmatrix}$ we can show that the matrices $a + b_j = \begin{pmatrix} a & -b \\ b & a \end{pmatrix}$ form a field under multiplication and addition which is isomorphic to $\mathbb{C}$ via $a + b_i \leftrightarrow \begin{pmatrix} a & -b \\ b & a \end{pmatrix}$.

Linear Combination

In mathematics, a linear combination is an expression constructed from a set of terms by multiplying each term by a constant and adding the results (e.g. a linear combination of x and y would be any expression of the form $ax + by$, where a and b are constants). The concept of linear combinations is central to linear algebra and related fields of mathematics. Most of this article deals with linear combinations in the context of a vector space over a field, with some generalizations given at the end of the article.

Definition

Suppose that K is a field (for example, the real numbers) and V is a vector space over K. As usual, we call elements of V *vectors* and call elements of K *scalars*. If $v_1,...,v_n$ are vectors and $a_1,...,a_n$ are scalars, then the *linear combination of those vectors with those scalars as coefficients* is

$$a_1\vec{v}_1 + a_2\vec{v}_2 + a_3\vec{v}_3 + \cdots + a_n\vec{v}_n.$$

There is some ambiguity in the use of the term "linear combination" as to whether it refers to the expression or to its value. In most cases the value is emphasized, like in the assertion "the set of all linear combinations of $v_1,...,v_n$ always forms a subspace". However, one could also say "two different linear combinations can have the same value" in which case the expression must have been meant. The subtle difference between these uses is the essence of the notion of linear dependence: a family F of vectors is linearly independent precisely if any linear combination of the vectors in F (as value) is uniquely so (as expression). In any case, even when viewed as expressions, all that matters about a linear combination is the coefficient of each v_i; trivial modifications such as permuting the terms or adding terms with zero coefficient do not give distinct linear combinations.

In a given situation, K and V may be specified explicitly, or they may be obvious from context. In that case, we often speak of *a linear combination of the vectors* $v_1,...,v_n$, with the coefficients unspecified (except that they must belong to K). Or, if S is a subset of V, we may speak of *a linear combination of vectors in S*, where both the coefficients and the vectors are unspecified, except that the vectors must belong to the set S (and the coefficients must belong to K). Finally, we may speak simply of *a linear combination*, where nothing is specified (except that the vectors must belong to *V* and the coefficients must belong to *K*); in this case one is probably referring to the expression, since every vector in *V* is certainly the value of some linear combination.

Note that by definition, a linear combination involves only finitely many vectors (except as described in Generalizations below). However, the set *S* that the vectors are taken from (if one is mentioned) can still be infinite; each individual linear combination will only involve finitely many vectors. Also, there is no reason that *n* cannot be zero; in that case, we declare by convention that the result of the linear combination is the zero vector in *V*.

Examples and Counterexamples

Euclidean Vectors

Let the field K be the set R of real numbers, and let the vector space V be the Euclidean space R^3. Consider the vectors $e_1 = (1,0,0)$, $e_2 = (0,1,0)$ and $e_3 = (0,0,1)$. Then any vector in R^3 is a linear combination of e_1, e_2 and e_3.

To see that this is so, take an arbitrary vector (a_1,a_2,a_3) in R^3, and write:

$$(a_1,a_2,a_3) = (a_1,0,0) + (0,a_2,0) + (0,0,a_3)$$

$$= a_1(1,0,0) + a_2(0,1,0) + a_3(0,0,1)$$

$$= a_1e_1 + a_2e_2 + a_3e_3.$$

Functions

Let *K* be the set C of all complex numbers, and let *V* be the set $C_C(R)$ of all continuous functions from the real line R to the complex plane C. Consider the vectors (functions) *f* and *g* defined by $f(t) := e^{it}$ and $g(t) := e^{-it}$. (Here, *e* is the base of the natural logarithm, about 2.71828..., and *i* is the imaginary unit, a square root of −1.) Some linear combinations of *f* and *g* are:

- $\cos t = \frac{1}{2}e^{it} + \frac{1}{2}e^{-it}$
- $2\sin t = (-i)e^{it} + (i)e^{-it}.$

On the other hand, the constant function 3 is *not* a linear combination of *f* and *g*. To see this, suppose that 3 could be written as a linear combination of e^{it} and e^{-it}. This means that there would exist complex scalars *a* and *b* such that $ae^{it} + be^{-it} = 3$ for all real numbers *t*. Setting $t = 0$ and $t = \pi$ gives the equations $a + b = 3$ and $a + b = -3$, and clearly this cannot happen.

Polynomials

Let K be R, C, or any field, and let V be the set P of all polynomials with coefficients taken from the field K. Consider the vectors (polynomials) $p_1 := 1$, $p_2 := x+1$, and $p_3 := x^2+x+1$.

Is the polynomial x^2-1 a linear combination of p_1, p_2, and p_3? To find out, consider an arbitrary linear combination of these vectors and try to see when it equals the desired vector x^2-1. Picking arbitrary coefficients a_1, a_2, and a_3, we want

$$a_1(1)+a_2(x+1)+a_3(x^2+x+1)=x^2-1.$$

Multiplying the polynomials out, this means

$$(a_1)+(a_2x+a_2)+(a_3x^2+a_3x+a_3)=x^2-1$$

and collecting like powers of x, we get

$$a_3x^2+(a_2+a_3)x+(a_1+a_2+a_3)=1x^2+0x+(-1).$$

Two polynomials are equal if and only if their corresponding coefficients are equal, so we can conclude

$$a_3=1, \quad a_2+a_3=0, \quad a_1+a_2+a_3=-1.$$

This system of linear equations can easily be solved. First, the first equation simply says that a_3 is 1. Knowing that, we can solve the second equation for a_2, which comes out to -1. Finally, the last equation tells us that a_1 is also -1. Therefore, the only possible way to get a linear combination is with these coefficients. Indeed,

$$x^2-1=-1-(x+1)+(x^2+x+1)=-p_1-p_2+p_3$$

so x^2-1 *is* a linear combination of p_1, p_2, and p_3.

On the other hand, what about the polynomial x^3-1? If we try to make this vector a linear combination of p_1, p_2, and p_3, then following the same process as before, we'll get the equation

$$0x^3+a_3x^2+(a_2+a_3)x+(a_1+a_2+a_3)$$

$$=1x^3+0x^2+0x+(-1).$$

However, when we set corresponding coefficients equal in this case, the equation for x^3 is

$$0=1$$

which is always false. Therefore, there is no way for this to work, and $x^3 - 1$ is *not* a linear combination of p_1, p_2, and p_3.

The Linear Span

Take an arbitrary field K, an arbitrary vector space V, and let $v_1,\ldots,v_n$ be vectors (in V). It's interesting to consider the set of *all* linear combinations of these vectors. This set is called the *linear span* (or just *span*) of the vectors, say S ={$v_1,\ldots,v_n$}. We write the span of S as span(S) or sp(S):

$$\mathrm{Sp}(v_1,\ldots,v_n) := \{a_1v_1 + \cdots + a_nv_n : a_1,\ldots,a_n \in K\}.$$

Linear Independence

For some sets of vectors $v_1,\ldots,v_n$, a single vector can be written in two different ways as a linear combination of them:

$$v = \sum a_i v_i = \sum b_i v_i \text{ where } (a_i) \neq (b_i).$$

Equivalently, by subtracting these ($c_i := a_i - b_i$) a non-trivial combination is zero:

$$0 = \sum c_i v_i.$$

If that is possible, then $v_1,\ldots,v_n$ are called *linearly dependent*; otherwise, they are *linearly independent*. Similarly, we can speak of linear dependence or independence of an arbitrary set S of vectors.

If S is linearly independent and the span of S equals V, then S is a basis for V.

Affine, Conical, and Convex Combinations

By restricting the coefficients used in linear combinations, one can define the related concepts of affine combination, conical combination, and convex combination, and the associated notions of sets closed under these operations.

Type of combination	Restrictions on coefficients	Name of set	Model space
Linear combination	no restrictions	Vector subspace	R^n
Affine combination	$\sum a_i = 1$	Affine subspace	Affine hyperplane
Conical combination	$a_i \geq 0$	Convex cone	Quadrant/Octant
Convex combination	$a_i \geq 0$ and $\sum a_i = 1$	Convex set	Simplex

Because these are more *restricted* operations, more subsets will be closed under them, so affine subsets, convex cones, and convex sets are *generalizations* of vector subspaces: a vector subspace is also an affine subspace, a convex cone, and a convex set, but a convex set need not be a vector subspace, affine, or a convex cone.

These concepts often arise when one can take certain linear combinations of objects, but not any: for example, probability distributions are closed under convex combination (they form a convex set), but not conical or affine combinations (or linear), and positive measures are closed under conical combination but not affine or linear – hence one defines signed measures as the linear closure.

Linear and affine combinations can be defined over any field (or ring), but conical and convex combination require a notion of "positive", and hence can only be defined over an ordered field (or ordered ring), generally the real numbers.

If one allows only scalar multiplication, not addition, one obtains a (not necessarily convex) cone; one often restricts the definition to only allowing multiplication by positive scalars.

All of these concepts are usually defined as subsets of an ambient vector space (except for affine spaces, which are also considered as "vector spaces forgetting the origin"), rather than being axiomatized independently.

Operad Theory

More abstractly, in the language of operad theory, one can consider vector spaces to be algebras over the operad R^{∞} (the infinite direct sum, so only finitely many terms are non-zero; this corresponds to only taking finite sums), which parametrizes linear combinations: the vector $(2,3,-5,0,\ldots)$ for instance corresponds to the linear combination $2v_1+3v_2-5v_3+0v_4+\cdots$. Similarly, one can consider affine combinations, conical combinations, and convex combinations to correspond to the sub-operads where the terms sum to 1, the terms are all non-negative, or both, respectively. Graphically, these are the infinite affine hyperplane, the infinite hyper-octant, and the infinite simplex. This formalizes what is meant by R^{n} being or the standard simplex being model spaces, and such observations as that every bounded convex polytope is the image of a simplex. Here suboperads correspond to more restricted operations and thus more general theories.

From this point of view, we can think of linear combinations as the most general sort of operation on a vector space – saying that a vector space is an algebra over the operad of linear combinations is precisely the statement that *all possible* algebraic operations in a vector space are linear combinations.

The basic operations of addition and scalar multiplication, together with the existence of an additive identity and additive inverses, cannot be combined in any more complicated way than the generic linear combination: the basic operations are a generating set for the operad of all linear combinations.

Ultimately, this fact lies at the heart of the usefulness of linear combinations in the study of vector spaces.

Generalizations

If V is a topological vector space, then there may be a way to make sense of certain *infinite* linear combinations, using the topology of V. For example, we might be able to speak of $a_1v_1 + a_2v_2 + a_3v_3 + ...$, going on forever. Such infinite linear combinations do not always make sense; we call them *convergent* when they do. Allowing more linear combinations in this case can also lead to a different concept of span, linear independence, and basis. The articles on the various flavours of topological vector spaces go into more detail about these.

If K is a commutative ring instead of a field, then everything that has been said above about linear combinations generalizes to this case without change. The only difference is that we call spaces like this V modules instead of vector spaces. If K is a noncommutative ring, then the concept still generalizes, with one caveat: Since modules over noncommutative rings come in left and right versions, our linear combinations may also come in either of these versions, whatever is appropriate for the given module. This is simply a matter of doing scalar multiplication on the correct side.

A more complicated twist comes when V is a bimodule over two rings, K_L and K_R. In that case, the most general linear combination looks like

$$a_1v_1b_1 + \cdots + a_nv_nb_n$$

where $a_1,...,a_n$ belong to K_L, $b_1,...,b_n$ belong to K_R, and $v_1,...,v_n$ belong to V.

Dual Space

In mathematics, any vector space V has a corresponding dual vector space (or just dual space for short) consisting of all linear functionals on V together with a naturally induced linear structure.

The dual space as defined above is defined for all vector spaces, and to avoid ambiguity may also be called the *algebraic dual space*. When defined for a topological vector space, there is a subspace of the dual space, corresponding to continuous linear functionals, called the *continuous dual space*.

Dual vector spaces find application in many branches of mathematics that use vector spaces, such as in tensor analysis with finite-dimensional vector spaces. When applied to vector spaces of functions (which are typically infinite-dimensional), dual spaces are used to describe measures, distributions, and Hilbert spaces. Consequently, the dual space is an important concept in functional analysis.

Algebraic Dual Space

Given any vector space V over a field F, the (algebraic) dual space V^* (alternatively denoted by $V^\vee$ or V') is defined as the set of all linear maps $\varphi: V \to F$ (linear functionals). Since linear maps are vector space homomorphisms, the dual space is also sometimes denoted by Hom(V, F). The dual space V^* itself becomes a vector space over F when equipped with an addition and scalar multiplication satisfying:

$$(\varphi + \psi)(x) = \varphi(x) + \psi(x)$$
$$(a\varphi)(x) = a\left(\varphi(x)\right)$$

for all φ and $\psi \in V^*$, $x \in V$, and $a \in F$. Elements of the algebraic dual space V^* are sometimes called covectors or one-forms.

The pairing of a functional φ in the dual space V^* and an element x of V is sometimes denoted by a bracket: $\varphi(x) = [x,\varphi]$ or $\varphi(x) = \langle\varphi,x\rangle$. This pairing defines a nondegenerate bilinear mapping $\langle\cdot,\cdot\rangle : V^* \times V \to F$ called the natural pairing.

Finite-dimensional Case

If V is finite-dimensional, then V^* has the same dimension as V. Given a basis $\{e_1, ..., e_n\}$ in V, it is possible to construct a specific basis in V^*, called the dual basis. This dual basis is a set $\{e^1, ..., e^n\}$ of linear functionals on V, defined by the relation

$$e^i(c^1e_1 + \cdots + c^ne_n) = c^i, \quad i = 1, \ldots, n$$

for any choice of coefficients $c^i \in F$. In particular, letting in turn each one of those coefficients be equal to one and the other coefficients zero, gives the system of equations

$$e^i(e_j) = \delta^i_j$$

where δ^i_j is the Kronecker delta symbol. For example, if V is R^2, and its basis chosen to be $\{e_1 = (1, 0), e_2 = (0, 1)\}$, then e^1 and e^2 are one-forms (functions that map a vector to a scalar) such that $e^1(e_1) = 1$, $e^1(e_2) = 0$, $e^2(e_1) = 0$, and $e^2(e_2) = 1$. (Note: The superscript here is the index, not an exponent).

In particular, if we interpret R^n as the space of columns of n real numbers, its dual space is typically written as the space of *rows* of n real numbers. Such a row acts on R^n as a linear functional by ordinary matrix multiplication. One way to see this is that a functional maps every n-vector x into a real number y. Then, seeing this functional as a matrix M, and x, y as a $n \times 1$ matrix and a 1×1 matrix (trivially, a real number) respectively, if we have $Mx = y$, then, by dimension reasons, M must be a $1 \times n$ matrix, i.e., M must be a row vector.

If V consists of the space of geometrical vectors in the plane, then the level curves of an element of V^* form a family of parallel lines in V, because the range is 1-dimensional, so that every point in the range is a multiple of any one nonzero element. So an element of V^* can be intuitively thought of as a particular family of parallel lines covering the plane. To compute the value of a functional on a given vector, one needs only to determine which of the lines the vector lies on. Or, informally, one "counts" how many lines the vector crosses. More generally, if V is a vector space of any dimension, then the level sets of a linear functional in V^* are parallel hyperplanes in V, and the action of a linear functional on a vector can be visualized in terms of these hyperplanes.

Infinite-dimensional Case

If V is not finite-dimensional but has a basis e_α indexed by an infinite set A, then the same construction as in the finite-dimensional case yields linearly independent elements e^α ($\alpha \in A$) of the dual space, but they will not form a basis.

Consider, for instance, the space R^∞, whose elements are those sequences of real numbers that contain only finitely many non-zero entries, which has a basis indexed by the natural numbers N: for $i \in N$, e_i is the sequence consisting of all zeroes except in the ith position, which is 1. The dual space of R^∞ is (isomorphic to) R^N, the space of all sequences of real numbers: such a sequence (a_n) is applied to an element (x_n) of R^∞ to give the number $\sum a_n x_n$, which is a finite sum because there are only finitely many nonzero x_n. The dimension of R^∞ is countably infinite, whereas R^N does not have a countable basis.

This observation generalizes to any infinite-dimensional vector space V over any field F: a choice of basis $\{e_\alpha : \alpha \in A\}$ identifies V with the space $(F^A)_0$ of functions $f : A \to F$ such that $f_\alpha = f(\alpha)$ is nonzero for only finitely many $\alpha \in A$, where such a function f is identified with the vector

$$\sum_{\alpha \in A} f_\alpha e_\alpha$$

in V (the sum is finite by the assumption on f, and any $v \in V$ may be written in this way by the definition of the basis).

The dual space of V may then be identified with the space F^A of *all* functions from A to F: a linear functional T on V is uniquely determined by the values $\theta_\alpha = T(e_\alpha)$ it takes on the basis of V, and any function $\theta : A \to F$ (with $\theta(\alpha) = \theta_\alpha$) defines a linear functional T on V by

$$T\Big(\sum_{\alpha \in A} f_\alpha e_\alpha\Big) = \sum_{\alpha \in A} f_\alpha T(e_\alpha) = \sum_{\alpha \in A} f_\alpha \theta_\alpha.$$

Again the sum is finite because f_α is nonzero for only finitely many α.

Note that $(F^A)_0$ may be identified (essentially by definition) with the direct sum of infinitely many copies of F (viewed as a 1-dimensional vector space over itself) indexed by A, i.e., there are linear isomorphisms

$$V \cong (F^A)_0 \cong \bigoplus_{\alpha \in A} F.$$

On the other hand, F^A is (again by definition), the direct product of infinitely many copies of F indexed by A, and so the identification

$$V^* \cong \Big(\bigoplus_{\alpha \in A} F\Big)^* \cong \prod_{\alpha \in A} F^* \cong \prod_{\alpha \in A} F \cong F^A$$

is a special case of a general result relating direct sums (of modules) to direct products.

Thus if the basis is infinite, then the algebraic dual space is *always* of larger dimension (as a cardinal number) than the original vector space. This is in contrast to the case of the continuous dual space, discussed below, which may be isomorphic to the original vector space even if the latter is infinite-dimensional.

Bilinear Products and dual Spaces

If V is finite-dimensional, then V is isomorphic to V^*. But there is in general no natural isomorphism between these two spaces. Any bilinear form $\langle\cdot,\cdot\rangle$ on V gives a mapping of V into its dual space via

$$v \mapsto \langle v, \cdot \rangle$$

where the right hand side is defined as the functional on V taking each $w \in V$ to $\langle v,w\rangle$. In other words, the bilinear form determines a linear mapping

$$\Phi_{\langle\cdot,\cdot\rangle} : V \to V$$

defined by

$$[\Phi_{\langle\cdot,\cdot\rangle}(v), w] = \langle v, w \rangle.$$

If the bilinear form is nondegenerate, then this is an isomorphism onto a subspace of V^*. If V is finite-dimensional, then this is an isomorphism onto all of V^*. Conversely, any isomorphism Φ from V to a subspace of V^* (resp., all of V^* if V is finite dimensional) defines a unique nondegenerate bilinear form $\langle\cdot,\cdot\rangle_\Phi$ on V by

$$\langle v, w \rangle_\Phi = (\Phi(v))(w) = [\Phi(v), w].$$

Thus there is a one-to-one correspondence between isomorphisms of V to subspaces of (resp., all of) V^* and nondegenerate bilinear forms on V.

If the vector space V is over the complex field, then sometimes it is more natural to consider sesquilinear forms instead of bilinear forms. In that case, a given sesquilinear form $\langle\cdot,\cdot\rangle$ determines an isomorphism of $\overline{V}$ with the complex conjugate of the dual space

$$\Phi_{\langle\cdot,\cdot\rangle} : V \to \overline{V^*}.$$

The conjugate space $\overline{V^*}$ can be identified with the set of all additive complex-valued functionals f: $V \to C$ such that

$$f(\alpha v) = \overline{\alpha} f(v).$$

Injection into the Double-dual

There is a natural homomorphism Ψ from V into the double dual $V^{**} = \{\Phi : V^* \to F : \Phi \text{ linear}\}$, defined by $(\Psi(v))(\varphi) = \varphi(v)$ for all $v \in V, \varphi \in V^*$. In other words, if $ev_v : V^* \to F$ is the evaluation map defined by $\varphi \mapsto \varphi(v)$, then we define $\Psi : V \to V^{**}$ as the map $v \mapsto ev_v$. This map Ψ is always injective; it is an isomorphism if and only if V is finite-dimensional. Indeed, the isomorphism of a finite-dimensional vector space with its double dual is an archetypal example of a natural isomorphism. Note that infinite-dimensional Hilbert spaces are not a counterexample to this, as they are isomorphic to their continuous duals, not to their algebraic duals.

Transpose of a Linear Map

If $f : V \to W$ is a linear map, then the *transpose* (or *dual*) $f^* : W^* \to V^*$ is defined by

$$f^*(\varphi) = \varphi \circ f$$

for every $\varphi \in W^*$. The resulting functional $f^*(\varphi)$ in V^* is called the *pullback* of φ along f.

The following identity holds for all $\varphi \in W^*$ and $v \in V$:

$$[f^*(\varphi), v] = [\varphi, f(v)],$$

where the bracket $[\cdot,\cdot]$ on the left is the natural pairing of V with its dual space, and that on the right is the natural pairing of W with its dual. This identity characterizes the transpose, and is formally similar to the definition of the adjoint.

The assignment $f \mapsto f^*$ produces an injective linear map between the space of linear operators from V to W and the space of linear operators from W^* to V^*; this homomorphism is an isomorphism if and only if W is finite-dimensional. If $V = W$ then the space of linear maps is actually an algebra under composition of maps, and the assignment is then an antihomomorphism of algebras, meaning that $(fg)^* = g^*f^*$. In the language of category theory, taking the dual of vector spaces and the transpose of linear maps is therefore a contravariant functor from the category of vector spaces over F to itself. Note that one can identify $(f^*)^*$ with f using the natural injection into the double dual.

If the linear map f is represented by the matrix A with respect to two bases of V and W, then f^* is represented by the transpose matrix A^T with respect to the dual bases of W^* and V^*, hence the name. Alternatively, as f is represented by A acting on the left on column vectors, f^* is represented by the same matrix acting on the right on row vectors. These points of view are related by the canonical inner product on R^n, which identifies the space of column vectors with the dual space of row vectors.

Quotient Spaces and Annihilators

Let S be a subset of V. The annihilator of S in V^*, denoted here S^o, is the collection of linear functionals $f \in V^*$ such that $[f, s] = 0$ for all $s \in S$. That is, S^o consists of all linear functionals $f: V \to F$ such that the restriction to S vanishes: $f|_S = 0$.

The annihilator of a subset is itself a vector space. In particular, $\emptyset^o = V^*$ is all of V^* (vacuously), whereas $V^o = 0$ is the zero subspace. Furthermore, the assignment of an annihilator to a subset of V reverses inclusions, so that if $S \subset T \subset V$, then

$$0 \subset T^o \subset S^o \subset V^*.$$

Moreover, if A and B are two subsets of V, then

$$(A \cap B)^o \supseteq A^o + B^o,$$

and equality holds provided V is finite-dimensional. If A_i is any family of subsets of V indexed by i belonging to some index set I, then

$$\left(\bigcup_{i \in I} A_i\right)^o = \bigcap_{i \in I} A_i^o.$$

In particular if A and B are subspaces of V, it follows that

$$(A + B)^o = A^o \cap B^o.$$

If V is finite-dimensional, and W is a vector subspace, then

$$W^{oo} = W$$

after identifying W with its image in the second dual space under the double duality isomorphism $V \approx V^{**}$. Thus, in particular, forming the annihilator is a Galois connection on the lattice of subsets of a finite-dimensional vector space.

If W is a subspace of V then the quotient space V/W is a vector space in its own right, and so has a dual. By the first isomorphism theorem, a functional $f : V \to F$ factors through V/W if and only if W is in the kernel of f. There is thus an isomorphism

$$(V / W)^* \cong W^o.$$

As a particular consequence, if V is a direct sum of two subspaces A and B, then V^* is a direct sum of A^o and B^o.

Continuous Dual Space

When dealing with topological vector spaces, one is typically only interested in the continuous linear functionals from the space into the base field $\mathbb{F} = \mathbb{C}$ (or $\mathbb{R}$). This gives rise to the notion of the "continuous dual space" or "topological dual" which is a linear subspace of the algebraic dual space V^*, denoted by V'. For any *finite-dimensional* normed vector space or topological vector space, such as Euclidean *n*-space, the continuous dual and the algebraic dual coincide. This is however false for any infinite-dimensional normed space, as shown by the example of discontinuous linear maps. Nevertheless, in the theory of topological vector spaces the terms "continuous dual space" and "topological dual space" are often replaced by "dual space", since there is no serious need to consider discontinuous maps in this field.

For a topological vector space V its *continuous dual space*, or *topological dual space*, or just *dual space* (in the sense of the theory of topological vector spaces) V' is defined as the space of all continuous linear functionals $\varphi : V \to \mathbb{F}$.

There is a standard construction for introducing a topology on the continuous dual V' of a topological vector space V. Fix a collection $\mathcal{A}$ of bounded subsets of V. Then one has the topology on V of uniform convergence on sets from $\mathcal{A}$, or what is the same thing, the topology generated by seminorms of the form

$$\| \varphi \|_A = \sup_{x \in A} | \varphi(x) |,$$

where φ is a continuous linear functional on V, and A runs over the class $\mathcal{A}$.

This means that a net of functionals φ_i tends to a functional φ in V' if and only if

$$\forall A \in \mathcal{A} \qquad \| \varphi_i - \varphi \|_A = \sup_{x \in A} | \varphi_i(x) - \varphi(x) | \underset{i \to \infty}{\longrightarrow} 0.$$

Usually (but not necessarily) the class $\mathcal{A}$ is supposed to satisfy the following conditions:

- each point x of V belongs to some set $A \in \mathcal{A}$

$$\forall x \in V \qquad \exists A \in \mathcal{A} \qquad x \in A,$$

- each two sets $A \in \mathcal{A}$ and $B \in \mathcal{A}$ are contained in some set $C \in \mathcal{A}$:

$$\forall A, B \in \mathcal{A} \qquad \exists C \in \mathcal{A} \qquad A \cup B \subseteq C,$$

- $\mathcal{A}$ is closed under the operation of multiplication by scalars:

$$\forall A \in \mathcal{A} \qquad \forall \lambda \in \mathbb{F} \qquad \lambda \cdot A \in \mathcal{A},$$

If these requirements are fulfilled then the corresponding topology on V' is Hausdorff and the sets

$$U_A = \{\varphi \in V' : \ \|\varphi\|_A < 1\}, \qquad A \in \mathcal{A},$$

form its local base.

Here are the three most important special cases.

- The strong topology on V' is the topology of uniform convergence on bounded subsets in V (so here $\mathcal{A}$ can be chosen as the class of all bounded subsets in V). If V is a normed vector space (e.g., a Banach space or a Hilbert space) then the strong topology on V' is normed (in fact a Banach space if the field of scalars is complete), with the norm

$$\|\varphi\| = \sup_{\|x\| \le 1} |\varphi(x)|.$$

- The stereotype topology on V' is the topology of uniform convergence on totally bounded sets in V (so here $\mathcal{A}$ can be chosen as the class of all totally bounded subsets in V).

- The weak topology on V' is the topology of uniform convergence on finite subsets in V (so here $\mathcal{A}$ can be chosen as the class of all finite subsets in V).

Each of these three choices of topology on V' leads to a variant of reflexivity property for topological vector spaces.

Examples

Let $1 < p < \infty$ be a real number and consider the Banach space ℓ^p of all sequences $a = (a_n)$ for which

$$\|a\|_p = \left(\sum_{n=0}^{\infty} |a_n|^p \right)^{1/p}$$

is finite. Define the number q by $1/p + 1/q = 1$. Then the continuous dual of ℓ^p is naturally identified with ℓ^q: given an element $\varphi \in (\ell^p)'$, the corresponding element of ℓ^q is the sequence ($\varphi(e_n)$) where e_n denotes the sequence whose n-th term is 1 and all others are zero. Conversely, given an element $a = (a_n) \in \ell^q$, the corresponding continuous linear functional φ on ℓ^p is defined by $\varphi(b) = \sum_n a_n b_n$ for all $b = (b_n) \in \ell^p$.

In a similar manner, the continuous dual of ℓ^1 is naturally identified with ℓ^∞ (the space of bounded sequences). Furthermore, the continuous duals of the Banach spaces c (consisting of all convergent sequences, with the supremum norm) and c_0 (the sequences converging to zero) are both naturally identified with ℓ^1.

By the Riesz representation theorem, the continuous dual of a Hilbert space is again a Hilbert space which is anti-isomorphic to the original space. This gives rise to the bra–ket notation used by physicists in the mathematical formulation of quantum mechanics.

By the Riesz–Markov–Kakutani representation theorem, the continuous dual of certain spaces of continuous functions can be described using measures.

Transpose of a Continuous Linear Map

If $T: V \to W$ is a continuous linear map between two topological vector spaces, then the (continuous) transpose $T': W' \to V'$ is defined by the same formula as before:

$$T'(\varphi) = \varphi \circ T, \quad \varphi \in W'.$$

The resulting functional $T'(\varphi)$ is in V'. The assignment $T \to T'$ produces a linear map between the space of continuous linear maps from V to W and the space of linear maps from W' to V'. When T and U are composable continuous linear maps, then

$$(U \circ T)' = T' \circ U'.$$

When V and W are normed spaces, the norm of the transpose in $L(W', V')$ is equal to that of T in $L(V, W)$. Several properties of transposition depend upon the Hahn–Banach theorem. For example, the bounded linear map T has dense range if and only if the transpose T' is injective.

When T is a compact linear map between two Banach spaces V and W, then the transpose T' is compact. This can be proved using the Arzelà–Ascoli theorem.

When V is a Hilbert space, there is an antilinear isomorphism i_V from V onto its continuous dual V'. For every bounded linear map T on V, the transpose and the adjoint operators are linked by

$$i_V \circ T^* = T' \circ i_V.$$

When T is a continuous linear map between two topological vector spaces V and W, then the transpose T' is continuous when W' and V' are equipped with"compatible" topologies: for example when, for $X = V$ and $X = W$, both duals X' have the strong topology $\beta(X', X)$ of uniform convergence on bounded sets of X, or both have the weak-* topology $\sigma(X', X)$ of pointwise convergence on X. The transpose T' is continuous from $\beta(W', W)$ to $\beta(V', V)$, or from $\sigma(W', W)$ to $\sigma(V', V)$.

Annihilators

Assume that W is a closed linear subspace of a normed space V, and consider the annihilator of W in V',

$$W^{\perp} = \{\varphi \in V' : W \subset \ker \varphi\}.$$

Then, the dual of the quotient V/W can be identified with $W^{\perp}$, and the dual of W can be identified with the quotient $V'/W^{\perp}$. Indeed, let P denote the canonical surjection from V onto the quotient V/W; then, the transpose P' is an isometric isomorphism from $(V/W)'$ into V', with range equal to $W^{\perp}$. If j denotes the injection map from W into V, then the kernel of the transpose j' is the annihilator of W:

$$\ker(j') = W^{\perp}$$

and it follows from the Hahn–Banach theorem that j' induces an isometric isomorphism $V'/W^{\perp} \to W'$.

Further Properties

If the dual of a normed space V is separable, then so is the space V itself. The converse is not true: for example the space ℓ^1 is separable, but its dual ℓ^∞ is not.

Topologies on the Dual

The topology of V and the topology of real or complex numbers can be used to induce on V' a dual space topology.

Double Dual

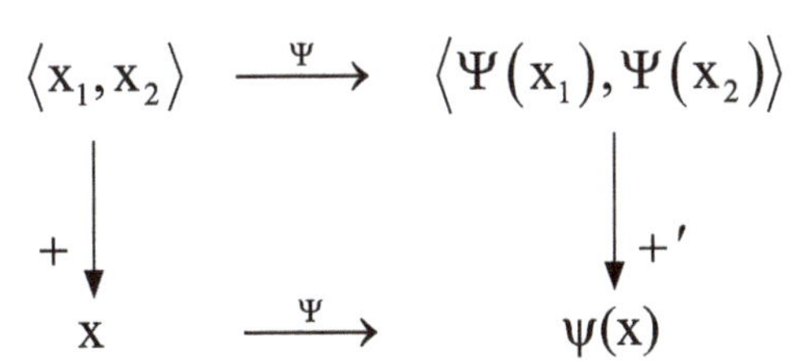

This is a natural transformation of vector addition from a vector space to its double dual. $\langle x_1, x_2\rangle$ denotes the ordered pair of two vectors. The addition + sends x_1 and x_2 to $x_1 + x_2$. The addition +′ induced by the transformation can be defined as $(\Psi(x_1) +' \Psi(x_2))(\varphi) = \varphi(x_1 + x_2) = \varphi(x)$ for any φ in the dual space.

In analogy with the case of the algebraic double dual, there is always a naturally defined continuous linear operator $\Psi : V \to V''$ from a normed space V into its continuous double dual V'', defined by

$$\Psi(x)(\varphi) = \varphi(x), \quad x \in V, \varphi \in V'.$$

As a consequence of the Hahn–Banach theorem, this map is in fact an isometry, meaning $\| \Psi(x) \| = \| x \|$ for all x in V. Normed spaces for which the map Ψ is a bijection are called reflexive.

When V is a topological vector space, one can still define $\Psi(x)$ by the same formula, for every $x \in V$, however several difficulties arise. First, when V is not locally convex, the continuous dual may be equal to {0} and the map Ψ trivial. However, if V is Hausdorff and locally convex, the map Ψ is injective from V to the algebraic dual V'^* of the continuous dual, again as a consequence of the Hahn–Banach theorem.

Second, even in the locally convex setting, several natural vector space topologies can be defined on the continuous dual V', so that the continuous double dual V'' is not uniquely defined as a set. Saying that Ψ maps from V to V'', or in other words, that $\Psi(x)$ is continuous on V' for every $x \in V$, is a reasonable minimal requirement on the topology of V', namely that the evaluation mappings

$$\varphi \in V' \mapsto \varphi(x), \quad x \in V,$$

be continuous for the chosen topology on V'. Further, there is still a choice of a topology on V'', and continuity of Ψ depends upon this choice. As a consequence, defining reflexivity in this framework is more involved than in the normed case.

Examples of Vector Spaces

We will let F denote an arbitrary field such as the real numbers R or the complex numbers C.

Trivial or Zero Vector Space

The simplest example of a vector space is the trivial one: {0}, which contains only the zero vector. Both vector addition and scalar multiplication are trivial. A basis for this vector space is the empty set, so that {0} is the 0-dimensional vector space over F. Every vector space over F contains a subspace isomorphic to this one.

The zero vector space is different from the null space of a linear operator F, which is the kernel of F.

Field

The next simplest example is the field F itself. Vector addition is just field addition and scalar multiplication is just field multiplication. Any non-zero element of F serves as a basis so F is a 1-dimensional vector space over itself.

The field is a rather special vector space; in fact it is the simplest example of a commutative algebra over F. Also, F has just two subspaces: {0} and F itself.

Coordinate Space

Perhaps the most important example of a vector space is the following. For any positive integer n, the space of all n-tuples of elements of F forms an n-dimensional vector space over F sometimes called coordinate space and denoted F^n. An element of F^n is written

$$x = (x_1, x_2, \ldots, x_n)$$

where each x_i is an element of F. The operations on F^n are defined by

$$x + y = (x_1 + y_1, x_2 + y_2, \ldots, x_n + y_n)$$

$$\alpha x = (\alpha x_1, \alpha x_2, \ldots, \alpha x_n)$$

$$0 = (0, 0, \ldots, 0)$$

$$-x = (-x_1, -x_2, \ldots, -x_n)$$

The most common cases are where F is the field of real numbers giving the real coordinate space R^n, or the field of complex numbers giving the complcx coordinate space C^n.

The quaternions and the octonions are respectively four- and eight-dimensional vector spaces over the reals.

The vector space F^n comes with a standard basis:

$$e_1 = (1, 0, \ldots, 0)$$

$$e_2 = (0, 1, \ldots, 0)$$

$$\vdots$$

$$e_n = (0, 0, \ldots, 1)$$

where 1 denotes the multiplicative identity in F.

Infinite Coordinate Space

Let F^∞ denote the space of infinite sequences of elements from F such that only *finitely* many elements are nonzero. That is, if we write an element of F^∞ as

$$x = (x_1, x_2, x_3, \ldots)$$

then only a finite number of the x_i are nonzero (i.e., the coordinates become all zero after a certain point). Addition and scalar multiplication are given as in finite coordinate space. The dimensionality of F^∞ is countably infinite. A standard basis consists of the vectors e_i which contain a 1 in the i-th slot and zeros elsewhere. This vector space is the coproduct (or direct sum) of countably many copies of the vector space F.

Note the role of the finiteness condition here. One could consider arbitrary sequences of elements in F, which also constitute a vector space with the same operations, often denoted by F^N - . F^N is the product of countably many copies of F.

By Zorn's lemma, F^N has a basis (there is no obvious basis). There are uncountably infinite elements in the basis. Since the dimensions are different, F^N is not isomorphic to F^∞. It is worth noting that F^N is (isomorphic to) the dual space of F^∞, because a linear map T from F^∞ to F is determined uniquely by its values $T(e_i)$ on the basis elements of F^∞, and these values can be arbitrary. Thus one sees that a vector space need not be isomorphic to its dual if it is infinite dimensional, in contrast to the finite dimensional case.

Product of Vector Spaces

Starting from *n* vector spaces, or a countably infinite collection of them, each with the same field, we can define the product space like above.

Matrices

Let $F^{m\times n}$ denote the set of m×n matrices with entries in F. Then $F^{m\times n}$ is a vector space over F. Vector addition is just matrix addition and scalar multiplication is defined in the obvious way (by multiplying each entry by the same scalar). The zero vector is just the zero matrix. The dimension of $F^{m\times n}$ is mn. One possible choice of basis is the matrices with a single entry equal to 1 and all other entries 0.

Polynomial Vector Spaces

One Variable

The set of polynomials with coefficients in F is a vector space over F, denoted F[x]. Vector addition and scalar multiplication are defined in the obvious manner. If the degree of the polynomials is unrestricted then the dimension of F[x] is countably infinite. If instead one restricts to polynomials with degree less than or equal to n, then we have a vector space with dimension $n + 1$.

One possible basis for F[x] is a monomial basis: the coordinates of a polynomial with respect to this basis are its coefficients, and the map sending a polynomial to the sequence of its coefficients is a linear isomorphism from F[x] to the infinite coordinate space F^∞.

The vector space of polynomials with real coefficients and degree less than or equal to n is denoted by P_n.

Several Variables

The set of polynomials in several variables with coefficients in F is vector space over F denoted $F[x_1, x_2, \ldots, x_r]$. Here r is the number of variables.

Function Spaces

Let X be a non-empty arbitrary set and V an arbitrary vector space over F. The space of all functions from X to V is a vector space over F under pointwise addition and multiplication. That is, let $f : X \to V$ and $g : X \to V$ denote two functions, and let $\alpha \in F$. We define

$$(f+g)(x) = f(x) + g(x)$$

$$(\alpha f)(x) = \alpha f(x)$$

where the operations on the right hand side are those in V. The zero vector is given by the constant function sending everything to the zero vector in V. The space of all functions from X to V is commonly denoted V^X.

If X is finite and V is finite-dimensional then V^X has dimension $|X|(\dim V)$, otherwise the space is infinite-dimensional (uncountably so if X is infinite).

Many of the vector spaces that arise in mathematics are subspaces of some function space. We give some further examples.

Generalized Coordinate Space

Let X be an arbitrary set. Consider the space of all functions from X to F which vanish on all but a finite number of points in X. This space is a vector subspace of F^X, the space of all possible functions from X to F. To see this, note that the union of two finite sets is finite, so that the sum of two functions in this space will still vanish outside a finite set.

The space described above is commonly denoted $(F^X)_0$ and is called generalized coordinate space for the following reason. If X is the set of numbers between 1 and n then this space is easily seen to be equivalent to the coordinate space F^n. Likewise, if X is the set of natural numbers, N, then this space is just F^∞.

A canonical basis for $(F^X)_0$ is the set of functions $\{\delta_x \mid x \in X\}$ defined by

$$\delta_x(y) = \begin{cases} 1 & x = y \\ 0 & x \neq y \end{cases}$$

The dimension of $(F^X)_0$ is therefore equal to the cardinality of X. In this manner we can construct a vector space of any dimension over any field. Furthermore, every vector space is isomorphic to one of this form. Any choice of basis determines an isomorphism by sending the basis onto the canonical one for $(F^X)_0$.

Generalized coordinate space may also be understood as the direct sum of $|X|$ copies of F (i.e. one for each point in X):

$$(F^X)_0 = \bigoplus_{x \in X} F.$$

The finiteness condition is built into the definition of the direct sum. Contrast this with the direct product of $|X|$ copies of F which would give the full function space F^X.

Linear Maps

An important example arising in the context of linear algebra itself is the vector space of linear maps. Let $L(V,W)$ denote the set of all linear maps from V to W (both of which are vector spaces over F). Then $L(V,W)$ is a subspace of W^V since it is closed under addition and scalar multiplication.

Note that $L(F^n,F^m)$ can be identified with the space of matrices $F^{m\times n}$ in a natural way. In fact, by choosing appropriate bases for finite-dimensional spaces V and W, L(V,W) can also be identified with $F^{m\times n}$. This identification normally depends on the choice of basis.

Continuous Functions

If X is some topological space, such as the unit interval [0,1], we can consider the space of all continuous functions from X to R. This is a vector subspace of R^X since the sum of any two continuous functions is continuous and scalar multiplication is continuous.

Differential Equations

The subset of the space of all functions from R to R consisting of (sufficiently differentiable) functions that satisfy a certain differential equation is a subspace of R^R if the equation is linear. This is because differentiation is a linear operation, i.e., (a f + b g)' = a f' + b g', where ' is the differentiation operator.

Field Extensions

Suppose K is a subfield of F (cf. field extension). Then F can be regarded as a vector space over K by restricting scalar multiplication to elements in K (vector addition is defined as normal). The dimension of this vector space is called the degree of the extension. For example the complex numbers C form a two-dimensional vector space over the real numbers R. Likewise, the real numbers R form an (uncountably) infinite-dimensional vector space over the rational numbers Q.

If V is a vector space over F it may also be regarded as vector space over K. The dimensions are related by the formula

$\dim_K V = (\dim_F V)(\dim_K F)$

For example C^n, regarded as a vector space over the reals, has dimension $2n$.

Finite Vector Spaces

Apart from the trivial case of a zero-dimensional space over any field, a vector space over a field F has a finite number of elements if and only if F is a finite field and the vector space has a finite dimension. Thus we have F_q, the unique finite field (up to isomorphism) with q elements. Here q must be a power of a prime ($q = p^m$ with p prime). Then any n-dimensional vector space V over F_q will have q^n elements. Note that the number of elements in V is also the power of a prime. The primary example of such a space is the coordinate space $(F_q)^n$.

Polynomials in one Variable

Any linear algebra can be used as a ring of scalars; we are mostly concerned with a ring $\mathbb{K}$ in general but the polynomial algebra $\mathbb{K}_\theta$ is also needed when the subject develops. We provide a discussion here. Recall that for any ring $\mathbb{K}$' two algebras $\mathbb{K}_\theta$ of polynomials as polynomial arise: one in which we write polynomial as $a(\theta) = \theta^n a_n + \ldots + \theta a_1 + a_0$, $a_i \varepsilon \mathbb{K}$, a_n and the other in which we write polynomial as $a(\theta) = a_n\theta^n + \ldots a_1\theta + a_0, a_i \varepsilon \mathbb{K}, a_n \neq 0$,

As modules, one of them is a right module and the other is a left module and the development runs in parallel. As rings they can be regarded as isomorphic via $f(a_n\theta^n) = \theta^n a_n$ the element $a_n \varepsilon\ \mathbb{K}$ is called the 'leading coefficient' of $a(\theta)$

(i) We take a ring $\mathbb{K}[\theta]$ of polynomial writing its elements as $a = a(\theta) = a_0 + \ldots + a_n\theta^n$, $a_n \neq 0$ and recall that it obey the following relations:

(a) If
$$a(\theta) = a_0 + a_1\theta + \ldots + a_n\theta^n, a_n \neq 0$$
$$b(\theta) = b_0 + b_1\theta + \ldots + b_m\theta^m, b_m \neq 0$$
$$c(\theta) = (a(\theta))(b(\theta)) = c_0 + c_1\theta + \ldots + = (ab)_0 + (ab)_1\theta + \ldots$$

then while this multiplication in $\mathbb{K}[\theta]$ is in general noncommutative

$\left[c_n = (ab)_n = \sum_{n=p+q} a_p b_q \neq \sum_{n=p+q} b_p a_q = (ba)_n \text{ in general}\right]$ we have $\deg[b(\theta)a(\theta)] = \deg[a(\theta)b(\theta)]$ $\left[= \deg(a(\theta)) + \deg(b(\theta)) = n + m\right]$ if either a_n or b_n is not a zero divisor in $\mathbb{K}$ i.e. if we do not get $a_n b_m = o \varepsilon K$ although $a_n \neq 0$, $b_m \neq 0$

note that this may happen if $\mathbb{K}$ is a matrix ring $Mat_n(\mathbb{L})$ for some ring $\mathbb{L}$

(b) deg[a(θ) + b(θ)] ≤ max{deg a(θ), deg b(θ)} = max{n, m}

(c) deg a(θ) = o iff a(θ) = $a_o \varepsilon \mathbb{K}$, we say that a = a(θ) is a constant.

(d) We set deg(o) = −∞ for the zero polynomial $o\varepsilon \mathbb{K}$ [θ].

(e) θ ε[θ] [i.e. θa(θ) = a(θ)θ for all a(θ)ε $\mathbb{K}$ [θ]] and thus each monomial {$\theta_k / \mathbb{K}$ ε $\mathbb{N}$ is in cen $\mathbb{K}$ [θ]}.

(ii) Proposition If μ(θ) has an invertible leading coefficient and deg a(θ) ≥ deg μ(θ), we can find some b(θ) with deg[a(θ) – μ(θ)b(θ)] < deg a(θ).

Proof: Say $a(\theta) = a_0 + \ldots + a_{n+m}\theta^{n+m}, \mu(\theta) = \mu_0 + \ldots + \mu_m\theta^m$

then with $b(\theta) = \mu_m^{-1} a_{m+n} \theta^n$, we have

$a(\theta) - \mu(\theta)b(\theta) = a_0 + \ldots + a_{n+m}\theta^{n+m} - \mu_m\mu_m^{-1}a_{n+m}\theta^{n+m} - \mu_{m-1}\mu_m^{-1}a_{n+m}\theta^{n+m-1} - \mu_0\mu_m^{-1}a_{n+m}\theta^n$

which has degree< n + m = deg a(θ) since the coefficient of θ^{n+m} is o.

We note that the same argument works to prove that we can find c(θ) with deg[a(θ)–c(θ) μ(θ)] < deg a(θ).

The Left Division Algorithm

Given a(θ)ε $\mathbb{K}$[θ], and μ(θ)ε $\mathbb{K}$[θ] with an invertible leading coefficient, there exists exactly one q(θ)εK[θ] such that a(θ) = μ(θ) + r(θ), deg r(θ) < deg μ(θ) and then r(θ) is also uniquely determined we say q is the 'quotient' and r is the remainder.

Proof: If there are two polynomials q(θ), $q'(\theta)$ satisfying the requirement so that a(θ) = μ(θ)q(θ) + r(θ) = μ(θ)qt(θ) + rt(θ), we have $\mu(\theta)[q(\theta) - q'(\theta)] = r(\theta) - r'(\theta)$

Assume $\deg\left[q(\theta) - q'(\theta)\right] = n$, deg μ(θ) = m so that LHS of (1) has degree n + m [∵ μ(θ) has invertible coefficient μ_m and thus if] λ is non zero we do not have $\mu_m \lambda = o$ since $\lambda = \mu_m^{-1}(\mu_m \lambda)$.

But deg(r(θ)) < deg μ(θ), $\deg(r'(\theta)) < \deg \mu(\theta)$ and $\deg\left[r'(\theta) - \deg r(\theta)\right] \leq \max\{\deg r'(\theta), \deg r(\theta)\} < \deg \mu(\theta)$ and so we have n + m < m which forces n = −∞ and thus $q(\theta) = q'(\theta)$ which then forces $r(\theta) = r'(\theta)$ and consequently uniqueness is established for q(θ) and then for r(θ) = a(θ) – μ(θ)q(θ).

For existence, we note that the set $S = \{\deg\left[a(\theta) - \mu(\theta)b(\theta)\right] | b(\theta) \in K[\theta]\}$ will have a least element; let q(θ) correspond to that i.e., let q(θ) be such that deg[a(θ) – μ(θ)qθ)]. We record r(θ) = a(θ) – μ(θ)q(θ); thus deg r(θ) is the least element of S. If deg r(θ) ≥ deg μ(θ) then we know there is some b(θ) with deg[r(θ) – μ(θ)b(θ)] < deg r(θ) so that with q_1 = q(θ) + b(θ), we have deg[a(θ) – μ(θ)q_1(θ)] < deg r(θ) which contradicts the choice of q(θ) that had assumed deg(a(θ) – μ(θ)q_1(θ)) as the least element of S. Therefore, we must have deg(r(θ)) < deg(μ(θ)) and we have found our q(θ) with a(θ) = μ(θ)q(θ) + r(θ), deg r(θ) < deg μ(θ) as required.

[this argument works for deg a(θθ) ≥ degμ(θ); if deg a(θ) < deg μ(θ) put q(θ) = o, r(θ) = a(θ). Further, note that if deg a(θ) = o, then degμ(θ) = o is forced since deg μ(θ) = ∞ is not permissible with invertible leading coefficient; then a(θ) = a_0, μ(θ) = μ_0 and μ_0 is invertible, take $q = \mu_0^{-1}a_0$].

(iii) In this preceding, q is the left quotient and r is the left remainder on left division by μ(θ); we say μ(θ) is a left divisor of a iff r=o, we similarly have

The Right Division Algorithm

Given a(θ) ε $\mathbb{K}$[θ], and μ(θ)ε $\mathbb{K}$[θ] with invertible leading coefficient, there is exactly one q(θ)ε $\mathbb{K}$[θ] such that a(θ) = q(θ)μ(θ) + r(θ) with deg r(θ) < deg μ(θ) and then r(θ) is also uniquely determined. Further, we say q(θ)is the right quotient, r(θ) is the right remainder on right division by μ(θ).

Example 1 If $a_3 = \begin{pmatrix} 1 & 2 \\ 0 & 0 \end{pmatrix}, \begin{pmatrix} -1 & 0 \\ 1 & 3 \end{pmatrix} a_1 = \begin{pmatrix} 2 & 3 \\ -2 & 0 \end{pmatrix}, a_0 = \begin{pmatrix} 0 & -1 \\ 1 & 2 \end{pmatrix}$ and $\mu_1 = \begin{pmatrix} 2 & 1 \\ 1 & 1 \end{pmatrix}$, $\mu_0 = \begin{pmatrix} 1 & -1 \\ 0 & 2 \end{pmatrix}$ where $a(\theta) = a_0 + a_1\theta + a_2\theta^2 + a_3\theta^3$, $\mu(\theta) = \mu_0 + \mu_1\theta$.

The right quotient and remainder are $q(\theta) = q_0 + q_1\theta + q_2\theta^2, r(\theta) = r_0$ with $q_2 = \begin{pmatrix} -1 & 3 \\ 0 & 0 \end{pmatrix}, q_1 = \begin{pmatrix} 7 & -14 \\ -2 & 5 \end{pmatrix}, q_0 = \begin{pmatrix} -43 & 81 \\ 12 & -24 \end{pmatrix}$ and $r_0 = \begin{pmatrix} 43 & -206 \\ 11 & 62 \end{pmatrix}$ while the left quotient is $q(\theta) = q_0 + q_1\theta + q_2\theta^2$ and the left remainder is $r_0 = \begin{pmatrix} -94 & -169 \\ 115 & 206 \end{pmatrix}$ on left and right division of a(θ) by μ(θ) respectively.

Proposition For a ring $\mathbb{K}$ and α, βε $\mathbb{K}$, the following are equivalent:

1. There is an invertible γε $\mathbb{K}$ such that $\beta = \gamma\alpha\gamma^{-1}$

2. There are invertibles p(θ), q(θ)ε $\mathbb{K}$ [θ] with θ – β = p(θ)(θ – α)q(θ)

Proof: (1) ⇒ (2) Since θεcen $\mathbb{K}$ [θ] and the constants γ, γ^{-1}ε $\mathbb{K}$ [θ], we find $\theta - \beta = \theta\gamma\gamma^{-1} - \gamma\alpha\gamma^{-1} = \gamma\theta\gamma^{-1} - \gamma\alpha\gamma^{-1} = \gamma(\theta - \alpha)\gamma^{-1}$ which proves (2).

(2) ⇒ (1)Let invertibles p(θ), q(θ)ε $\mathbb{K}$ [θ] be given with θ – β = p(θ)(θ – α)q(θ). Since θ – α has leading coefficient $1 \neq 0$ the division algorithm assume the existence of uniquely given a(θ)and u(θ)such that (p(θ)$^{-1}$)ε $\mathbb{K}$ [θ] can be written as (p(θ)$^{-1}$) = (θ – α)a(θ) + u(θ), deg u(θ) < deg(θ – α) = 1 which forces u = u(θ)ε $\mathbb{K}$, then u(θ – β) = [p(θ)$^{-1}$ – (θ – α)a(θ)] (θ – β) = (p(θ))–1(θ – β) – (θ – α)a(θ) (θ – β) = (p(θ)) p(θ)(θ – α) q(θ) – (θ – α) a(θ)(θ – β) = (θ – α) [q(θ) – a(θ)(θ – β)]

Comparing the highest degree terms on both sides, we get q(θ) – a(θ)(θ – β)u. The equation is thus u(θ – β) = (θ – α)u which has forces uβ = αu. Dividing p(θ) by (θ – β) [which has leading coefficient $1 \neq 0$] we get p(θ) = (θ – β)b(θ) + r(θ), with deg r(θ) < deg (θ – α) = 1 forcing r(θ) = rε $\mathbb{K}$.

We have 1 = p(θ)[p(θ)]–1 = p(θ)[(θ – α)a(θ) + u] = p(θ)(θ – α)a(θ) + p(θ)u = (θ – β) [q(θ)]$^{-1}$a(θ) + [(θ – β)p(θ + r)]u = (θ – β) [q(θ)a(θ) + b(0)u] + ru so that 1 – ru = (θ – β) [q(θ) a(θ) + b(θ)u]. Comparing the coefficient on both sides, we get RHS=0 hence 1 – ru = 0 which forces ru = 1. Further, we have 1 = [p(θ)]$^{-1}$p(θ) = [(θ – α)a(θ)+ u]p(θ) = [(θ – α)a (θ)+ u] [(θ – β) b(θ)+ r] = (θ – α) a(θ) (θ – β) b(θ)+(θ – α) a(θ)r + u(θ – β) b(θ)+ ur so that 1 – ur = (θ – α) a(θ)(θ – β) b(θ)+ (θ – α)a (θ)r + u(θ – β)b(θ)Comparing the coefficients of θ on both sides, we get RHS=0 which forces 1 – ur = 0. Therefore, u is invertible with inverse r.

Euclidean Domain

In mathematics, more specifically in abstract algebra and ring theory, a Euclidean domain (also called a Euclidean ring) is a commutative ring that can be endowed with a Euclidean function (explained below) which allows a suitable generalization of the Euclidean division of the integers. This generalized Euclidean algorithm can be put to many of the same uses as Euclid's original algorithm in the ring of integers: in any Euclidean domain, one can apply the Euclidean algorithm to compute the greatest common divisor of any two elements. In particular, the greatest common divisor of any two elements exists and can be written as a linear combination of them (Bézout's identity). Also every ideal in a Euclidean domain is principal, which implies a suitable generalization of the fundamental theorem of arithmetic: every Euclidean domain is a unique factorization domain.

It is important to compare the class of Euclidean domains with the larger class of principal ideal domains (PIDs). An arbitrary PID has much the same "structural properties" of a Euclidean domain (or, indeed, even of the ring of integers), but when an explicit algorithm for Euclidean division is known, one may use Euclidean algorithm and extended Euclidean algorithm to compute greatest common divisors and Bézout's identity. In particular, the existence of efficient algorithms for Euclidean division of integers and of polynomials in one variable over a field is of basic importance in computer algebra.

So, given an integral domain *R*, it is often very useful to know that *R* has a Euclidean function: in particular, this implies that *R* is a PID. However, if there is no "obvious" Euclidean function, then determining whether R is a PID is generally a much easier problem than determining whether it is a Euclidean domain.

Euclidean domains appear in the following chain of class inclusions:

> commutative rings ⊃ integral domains ⊃ integrally closed domains ⊃ GCD domains ⊃ unique factorization domains ⊃ principal ideal domains ⊃ Euclidean domains ⊃ fields ⊃ finite fields

Definition

Let R be an integral domain. A Euclidean function on R is a function f from $R\setminus\{0\}$ to the non-negative integers satisfying the following fundamental division-with-remainder property:

- (EF1) If a and b are in R and b is nonzero, then there are q and r in R such that $a = bq + r$ and either $r = 0$ or $f(r) < f(b)$.

A Euclidean domain is an integral domain which can be endowed with at least one Euclidean function. It is important to note that a particular Euclidean function f is *not* part of the structure of a Euclidean domain: in general, a Euclidean domain will admit many different Euclidean functions.

Most algebra texts require a Euclidean function to have the following additional property:

- (EF2) For all nonzero a and b in R, $f(a) \leq f(ab)$.

However, one can show that (EF2) is superfluous in the following sense: any domain R which can be endowed with a function g satisfying (EF1) can also be endowed with a function f satisfying (EF1) and (EF2): indeed, for $a \in R\setminus\{0\}$ one can define $f(a)$ as follows

$$f(a) = \min_{x\in R\setminus\{0\}} g(xa)$$

In words, one may define $f(a)$ to be the minimum value attained by g on the set of all non-zero elements of the principal ideal generated by a.

A multiplicative Euclidean function is one such that $f(ab)=f(a)f(b)$ and $f(a)$ is never zero. It follows that $f(1)=1$ and in fact $f(a)=1$ if and only if a is a unit.

Notes on the Definition

Many authors use other terms such as "degree function", "valuation function", "gauge function" or "norm function", in place of "Euclidean function". Some authors also require the domain of the Euclidean function to be the entire ring R; however this does not essentially affect the definition, since (EF1) does not involve the value of $f(0)$. The definition is sometimes generalized by allowing the Euclidean function to take its values in any well-ordered set; this weakening does not affect the most important implications of the Euclidean property.

The property (EF1) can be restated as follows: for any principal ideal I of R with nonzero generator b, all nonzero classes of the quotient ring R/I have a representative r with $f(r) < f(b)$. Since the possible values of f are well-ordered, this property can be established by proving $f(r) < f(b)$ for any r (not in I) with minimal value of $f(r)$ in its class. Note that for a Euclidean function that is so established there need not exist an effective method to determine q and r in (EF1).

Examples

Examples of Euclidean domains include:

- Any field. Define $f(x) = 1$ for all nonzero x.
- Z, the ring of integers. Define $f(n) = |n|$, the absolute value of n.
- Z[i], the ring of Gaussian integers. Define $f(a + bi) = a^2 + b^2$, the squared norm of the Gaussian integer $a + bi$.
- Z[ω] (where ω is a primitive (non-real) cube root of unity), the ring of Eisenstein integers. Define $f(a + b\omega) = a^2 - ab + b^2$, the norm of the Eisenstein integer $a + b\omega$.
- $K[X]$, the ring of polynomials over a field K. For each nonzero polynomial P, define $f(P)$ to be the degree of P.
- $K[[X]]$, the ring of formal power series over the field K. For each nonzero power series P, define $f(P)$ as the degree of the smallest power of X occurring in P. In particular, for two nonzero power series P and Q, $f(P) \le f(Q)$ iff P divides Q.
- Any discrete valuation ring. Define $f(x)$ to be the highest power of the maximal ideal M containing x (equivalently, to the power of the generator of the maximal ideal that x is associated to). The previous case $K[[X]]$ is a special case of this.
- A Dedekind domain with finitely many nonzero prime ideals $P_1, \ldots, P_n$. Define $f(x) = \sum_{i=1}^{n} v_i(x)$, where v_i is the discrete valuation corresponding to the ideal P_i. (Samuel 1971)

Example of domains that are not Euclidean domains include

- Every domain that is not a principal ideal domain, such as the ring of polynomials in at least two indeterminates over a field, or the ring of univariate polynomials with integer coefficients

- The ring of integers of $\mathbb{Q}(\sqrt{-19})$, consisting of the numbers $\frac{a+b\sqrt{-19}}{2}$ such that a and b are integers, that are either both even or both odd. It is a principal ideal domain that is not Euclidean.
- The ring $\mathbb{R}[X,Y]/(X^2+Y^2+1)$ is also a principal ideal domain that is not Euclidean.

Properties

Let R be a domain and f a Euclidean function on R. Then:

- R is a principal ideal domain (PID). In fact, if I is a nonzero ideal of R then any element a of $I\setminus\{0\}$ with minimal value (on that set) of $f(a)$ is a generator of I. As a consequence R is also a unique factorization domain and a Noetherian ring. With respect to general principal ideal domains, the existence of factorizations (i.e., that R is an atomic domain) is particularly easy to prove in Euclidean domains: choosing a Euclidean function f satisfying (EF2), x cannot have any decomposition into more than $f(x)$ nonunit factors, so starting with x and repeatedly decomposing reducible factors is bound to produce a factorization into irreducible elements.
- Any element of R at which f takes its globally minimal value is invertible in R. If an f satisfying (EF2) is chosen, then the converse also holds, and f takes its minimal value exactly at the invertible elements of R.
- If the Euclidean property is algorithmic, i.e., if there is a division algorithm that for given a and nonzero b produces a quotient q and remainder r with $a = bq + r$ and either $r = 0$ or $f(r) < f(b)$, then an extended Euclidean algorithm can be defined in terms of this division operation.
- If a Euclidean domain is not a field then it has an element a with the following property: any element x not divisible by a can be written as $x=ay+u$ for some unit u and some element y. This follows by taking a to be a non-unit with $f(a)$ as small as possible. This strange property can be used to show that some principal ideal domains are not Euclidean domains, as not all PIDs have this property. For example, for $d = -19, -43, -67, -163$, the ring of integers of $\mathbb{Q}(\sqrt{d})$ is a PID which is not Euclidean, but the cases $d = -1, -2, -3, -7, -11$ are Euclidean.

However, in many finite extensions of Q with trivial class group, the ring of integers is Euclidean (not necessarily with respect to the absolute value of the field norm. Assuming the extended Riemann hypothesis, if K is a finite extension of Q and the ring of integers of K is a PID with an infinite number of units, then the ring of integers is Euclidean. In particular this applies to the case of totally real quadratic number fields with trivial class group. In addition (and without assuming ERH), if the field K is a Galois extension of Q, has trivial class group and unit rank strictly greater than three, then the ring of integers is Euclidean. An immediate corollary of this is that if the number field is Galois over Q, its class group is trivial and the extension has degree greater than 8 then the ring of integers is necessarily Euclidean.

Norm-Euclidean Fields

Algebraic number fields K come with a canonical norm function on them: the absolute value of the field norm N that takes an algebraic element α to the product of all the conjugates of α. This norm maps the ring of integers of a number field K, say O_K, to the nonnegative rational integers, so it is a candidate to be a Euclidean norm on this ring. If this norm satisfies the axioms of a Euclidean function then the number field K is called *norm-Euclidean* or simply *Euclidean*. Strictly speaking it is the ring of integers that is Euclidean since fields are trivially Euclidean domains, but the terminology is standard.

If a field is not norm-Euclidean then that does not mean the ring of integers is not Euclidean, just that the field norm does not satisfy the axioms of a Euclidean function. In fact, the rings of integers of number fields may be divided in several classes:

- Those that are not principal and therefore not Euclidean, such as the integers of $\mathbb{Q}(\sqrt{-5})$
- Those that are principal and not Euclidean, such as the integers of $\mathbb{Q}(\sqrt{-19})$
- Those that are Euclidean and not norm-Euclidean, such as the integers of $\mathbb{Q}(\sqrt{69})$
- Those that are norm-Euclidean, such as Gaussian integers (integers of $\mathbb{Q}(\sqrt{-1})$)

The norm-Euclidean quadratic fields have been fully classified, they are $\mathbb{Q}(\sqrt{d})$ where d takes the values

−11, −7, −3, −2, −1, 2, 3, 5, 6, 7, 11, 13, 17, 19, 21, 29, 33, 37, 41, 57, 73 (sequence A048981 in the OEIS).

Every Euclidean imaginary quadratic field is norm-Euclidean and is one of the five first fields in the preceding list.

We say a ring is an entire ring iff $\lambda\mu = 0 \Rightarrow$ either $\lambda = 0$ or $\mu = 0$; a commutative entire ring is also called integral domain. An integral domain $\mathbb{K}$ is called a Euclidean domain iff there is an 'Euclidean function' $\mathbb{K}\big|\{0\} \rightarrow^{g} \mathbb{N}$ satisfying

E_1 if λ divides $\mu[\neq 0]$ then $g(\lambda) \leq g(\mu)$

E_2 For every pair of elements α, β of $\mathbb{K}$, $\alpha \neq 0$

there exists elements $\gamma, \delta \varepsilon \mathbb{K}$ with β = αγ + δ with δ = o or g(δ) < g(α). [thus in a domain, it is called 'Euclidean algorithm']. In the preceding section, we proved that if $\mathbb{F}$ is a field, $\mathbb{F}[\theta]$ is a Euclidean domain [g(a(θ)) = deg a(θ)] The ring of integers $\mathbb{Z}$ is also an integral domain [This is the reason for 'integral' domain'] with g(a) = |a|, aε $\mathbb{Z}$. Indeed, if $b = ac \neq 0$ then |c| ≥ 1 and hence |b| = |c||a| ≥ |a|. Further, for any two integers $a, b, a \neq 0$ the division algorithm in $\mathbb{Z}$ ensures b = |a|q + r = a(±q) + r with r = o or o < r < |a|. These two examples of Euclidean domains are the ones we shall use in this course.

However, there are other Euclidean domains: let $d \neq 1$ be a square free integer [in the sense that its prime factorization has no square] and let $\theta = \begin{cases} \frac{1+\sqrt{d}}{2} & \text{if } d \equiv 1 \bmod 4 \\ \sqrt{d} & \text{otherwise} \end{cases}$.

Consider $\mathbb{Q}\left[\sqrt{d}\right]$ which is a field; for $\alpha = p + q\sqrt{d}$ we write $p^2 - q^2 d$ as N (α). Then for d = −1, −2, −3, −7, −11, 2, 3, 5, 6, 7, 13, 17, 21, 29, the function g(α) = |N (α)| is a Euclidean function; we shall not prove this here.

Proposition: The gcd of any two elements α, β not both zero, of Euclidean domain $\mathbb{E}$ exists and can be expressed as αλ + βμ with λ, με $\mathbb{E}$.

Proof: Suppose g is the Euclidean function, g(α) ≥ g(β). Then by the property of g [the 'Euclidean algorithm', also called 'division algorithm'] we have $\alpha = \beta\gamma_1 + \delta_1$, $g(\delta_1) < g(\beta)$; $\beta = \delta_1\gamma_2 + \delta_2$, $g(\delta_2) < g(\delta_1)$...

Thus $g(\beta) > g(\delta_1) > g(\delta_2)$... is a decreasing sequence of nonnegative integers which must stop and after some time, we have $\delta_{n-2} = \delta_{n-1}\gamma_n + \delta_n$, $\delta_{n-1} = \delta_n\gamma_{n+1} + \delta_{n+1}$ with $\delta_{n+1} = o = \delta_k$ for any k > n + 1 now $\delta_1 = \alpha - \beta\gamma_1$ so δ_1 has the form αλ + βμ with λ = 1, $\mu = -\gamma_1$. In general, if $\delta_{i-1} = \alpha\lambda_{i-1} + \beta\mu_{i-1}$ and $\delta_{i-2} = \alpha\lambda_{i-2} + \beta\mu_{i-2}$

$\delta_i = -\delta_{i-1}\gamma_i + \delta_{i-2} = -(\alpha\lambda_{i-1} + \beta\mu_{i-1})\gamma_i + \alpha\lambda_{i-2} + \beta\mu_{i-2} = \alpha[\lambda_{i-2} - \lambda_{i-1}] + \beta[\mu_{i-2} - \mu_{i-1}\gamma_i]$ also has this form αλ + βμ [with $\lambda = \lambda_{i-2} - \lambda_{i-1}$, $\mu = \mu_{i-2} - \mu_{i-1}\gamma_i$]. Thus $\delta_n = \alpha\lambda_n + \beta\mu_n$ for some λ_n, $\mu_n \varepsilon\, \mathbb{E}$. Now $\delta_n = \delta_n.1 + o$ so δ_n divides δ_n and $\delta_{n-1} = \delta_n\gamma_{n+1} + o$ so δ_n divides δ_{n-1}. But we have $\delta_{n-2} = \delta_{n-1}\gamma_n + \delta_n = \delta_n\gamma_{n+1}\gamma_n + \delta_n = \delta_n[\gamma_{n+1}\gamma_{n+1}]$ so δ_n divides δ_{n-2}. Similarly, δ_n divides all the 'remainders' δ_i. Let $\delta_i = \delta_n p_i$, $p_i \varepsilon\, \mathbb{E}$. Then $\beta = \delta_1\gamma_2 + \delta_2 = \delta_n p_1\gamma_2 + \delta_n p_2 = \delta_n[p_1\gamma_2 + \beta_2]$ so that δ_n divides β and $\alpha = \beta\gamma_1 + \delta_1 = \delta_n[p_1\gamma_2 + \beta_2]\gamma_1 + \delta_n p_1 = \delta_n[p_1\gamma_2\gamma_1 + p_1]$ so that δ_n divides α.

Thus is a common divisor of α and β. If γ is any other divisor of α and β say α = γp, β = γq, p, qε $\mathbb{E}$ then $\delta_n = \alpha\lambda_n + \beta\mu_n = \gamma_p\lambda_n + \gamma_q\mu_n = \gamma[p\lambda_n + q\mu_n]$ so that γ divides δ_n. Thus δ_n is the greatest common divisor of α and β, and is of the form αλ + βμ [with $\lambda = \lambda_n$, $\mu = \mu_n$]. This proves that advertized result. But there is a further piece of information. Suppose $\alpha = \delta_n a$, $\beta = \delta_n b$ then writing δ_n m = αβ, we have $\delta_n m = \delta_n a \delta_n b = a\delta_n^2 b = \delta_n^2 ab$ [because of commutativity] i.e. m = aβ and m = bα [∵ $\mathbb{E}$ is a integral domain, $\delta_n \neq 0$] so that m is a common multiple. If m' is another common multiple, then $m' = \alpha c_1, m' = \beta c_2$ hence $\delta_n m' = (\alpha\lambda_n + \beta\mu_n)m' = \alpha m'\lambda_n + \beta m'\mu_n = \alpha\beta c_2\lambda_n + \beta\alpha c_1\mu_n = \alpha\beta[c_2\lambda_n + c_1\mu_n] = \delta_n m[c_2\lambda_n + c_1\mu_n]$

But then $\delta_n\left[m'-m\left(c_2\lambda_n+c_1\mu_n\right)\right]=0$ and since $\delta_n\neq 0$, and E is an integral domain, we get $m'-m\left(c_2\lambda_n+c_1\mu_n\right)=0$ which means $m'=m\left(c_2\lambda_n+c_1\mu_n\right)$ so that m is a divisor of m'. Thus m is the least common multiple of α, β.

To sum up: In a Euclidean domain, both the LCM and GCD exists.

Now suppose A is an ideal in a Euclidean domain $\mathbb{E}$. Pick aεA with a = bq + r where g(b) has the least value for element of A. Then since a, bεA, we have r = a – bqεA with g(r) < g(b) which is not possible since b has the least value g(b) for elements of A. Thus r = 0 and we have A = (a) = {aλ|λ ∈ $\mathbb{E}$}[1]. Such an ideal which is generated by a single element is called a principal ideal. To sum up: Every Euclidean domain $\mathbb{E}$ is a principal ideal domain in the sense that every ideal in $\mathbb{E}$ is a principal ideal.

Linear Subspace

In linear algebra and related fields of mathematics, a linear subspace, also known as a vector subspace, or, in the older literature, a linear manifold, is a vector space that is a subset of some other (higher-dimension) vector space. A linear subspace is usually called simply a *subspace* when the context serves to distinguish it from other kinds of subspaces.

Definition

Let *K* be a field (such as the real numbers), *V* be a vector space over *K*, and let *W* be a subset of *V*. Then *W* is a subspace if:

1. The zero vector, 0, is in *W*.
2. If u and v are elements of W, then the sum u + v is an element of W.
3. If u is an element of *W* and c is a scalar from *K*, then the scalar product cu is an element of *W*.

Examples

Example I: Let the field K be the set R of real numbers, and let the vector space V be the real coordinate space R^3. Take *W* to be the set of all vectors in *V* whose last component is 0. Then *W* is a subspace of *V*.

Proof:

1. Given u and v in *W*, then they can be expressed as u = $(u_1, u_2, 0)$ and v = $(v_1, v_2, 0)$. Then u + v = $(u_1+v_1, u_2+v_2, 0+0)$ = $(u_1+v_1, u_2+v_2, 0)$. Thus, u + v is an element of *W*, too.

2. Given u in W and a scalar c in R, if u = $(u_1, u_2, 0)$ again, then cu = $(cu_1, cu_2, c0)$ = $(cu_1, cu_2, 0)$. Thus, cu is an element of W too.

Example II: Let the field be R again, but now let the vector space be the Cartesian plane R^2. Take W to be the set of points (x, y) of R^2 such that x = y. Then W is a subspace of R^2.

Proof:

1. Let p = (p_1, p_2) and q = (q_1, q_2) be elements of W, that is, points in the plane such that $p_1 = p_2$ and $q_1 = q_2$. Then p + q = (p_1+q_1, p_2+q_2); since $p_1 = p_2$ and $q_1 = q_2$, then $p_1 + q_1 = p_2 + q_2$, so p + q is an element of W.
2. Let p = (p_1, p_2) be an element of W, that is, a point in the plane such that $p_1 = p_2$, and let c be a scalar in R. Then cp = (cp_1, cp_2); since $p_1 = p_2$, then $cp_1 = cp_2$, so cp is an element of W.

In general, any subset of the real coordinate space R^n that is defined by a system of homogeneous linear equations will yield a subspace. (The equation in example I was $z = 0$, and the equation in example II was $x = y$.) Geometrically, these subspaces are points, lines, planes, and so on, that pass through the point 0.

Examples Related to Calculus

Example III: Again take the field to be R, but now let the vector space V be the set R^R of all functions from R to R. Let C(R) be the subset consisting of continuous functions. Then C(R) is a subspace of R^R.

Proof:

1. We know from calculus that $0 \in C(R) \subset R^R$.
2. We know from calculus that the sum of continuous functions is continuous.
3. Again, we know from calculus that the product of a continuous function and a number is continuous.

Example IV: Keep the same field and vector space as before, but now consider the set Diff(R) of all differentiable functions. The same sort of argument as before shows that this is a subspace too.

Examples that extend these themes are common in functional analysis.

Properties of Subspaces

A way to characterize subspaces is that they are closed under linear combinations. That is, a nonempty set W is a subspace if and only if every linear combination of (finitely many) elements of W also belongs to W. Conditions 2 and 3 for a subspace are simply the most basic kinds of linear combinations.

In a topological vector space X, a subspace W need not be closed in general, but a finite-dimensional subspace is always closed. The same is true for subspaces of finite codimension, i.e. determined by a finite number of continuous linear functionals.

Descriptions

Descriptions of subspaces include the solution set to a homogeneous system of linear equations, the subset of Euclidean space described by a system of homogeneous linear parametric equations, the span of a collection of vectors, and the null space, column space, and row space of a matrix. Geometrically (especially, over the field of real numbers and its subfields), a subspace is a flat in an n-space that passes through the origin.

A natural description of an 1-subspace is the scalar multiplication of one non-zero vector v to all possible scalar values. 1-subspaces specified by two vectors are equal if and only if one vector can be obtained from another with scalar multiplication:

$$\exists c \in K : v' = cv \ (\text{or } v = \frac{1}{c} v')$$

This idea is generalized for higher dimensions with linear span, but criteria for equality of k-spaces specified by sets of k vectors are not so simple.

A dual description is provided with linear functionals (usually implemented as linear equations). One non-zero linear functional F specifies its kernel subspace F = 0 of codimension 1. Subspaces of codimension 1 specified by two linear functionals are equal if and only if one functional can be obtained from another with scalar multiplication (in the dual space):

$$\exists c \in K : F' = cF \ (\text{or } F = \frac{1}{c} F')$$

It is generalized for higher codimensions with a system of equations. The following two subsections will present this latter description in details, and the remaining four subsections further describe the idea of linear span.

Systems of Linear Equations

The solution set to any homogeneous system of linear equations with n variables is a subspace in the coordinate space K^n:

$$\left\{ \begin{bmatrix} x_1 \\ x_2 \\ \vdots \\ x_n \end{bmatrix} \in K^n : \begin{matrix} a_{11}x_1 + a_{12}x_2 + \ldots + a_{1n}x_n = 0 \\ a_{21}x_1 + a_{22}x_2 + \ldots + a_{2n}x_n = 0 \\ \vdots \\ a_{m1}x_1 + a_{m2}x_2 + \ldots + a_{mn}x_n = 0 \end{matrix} \right\}.$$

For example (over real or rational numbers), the set of all vectors (x, y, z) satisfying the equations

$$x + 3y + 2z = 0 \quad \text{and} \quad 2x - 4y + 5z = 0$$

is a one-dimensional subspace. More generally, that is to say that given a set of n independent functions, the dimension of the subspace in K^k will be the dimension of the null set of A, the composite matrix of the n functions.

Null Space of a Matrix

In a finite-dimensional space, a homogeneous system of linear equations can be written as a single matrix equation:

$$Ax = 0.$$

The set of solutions to this equation is known as the null space of the matrix. For example, the subspace described above is the null space of the matrix

$$A = \begin{bmatrix} 1 & 3 & 2 \\ 2 & -4 & 5 \end{bmatrix}.$$

Every subspace of K^n can be described as the null space of some matrix.

Linear Parametric Equations

The subset of K^n described by a system of homogeneous linear parametric equations is a subspace:

$$\left\{ \begin{bmatrix} x_1 \\ x_2 \\ \vdots \\ x_n \end{bmatrix} \in K^n : \begin{matrix} x_1 = a_{11}t_1 + a_{12}t_2 + \ldots + a_{1m}t_m \\ x_2 = a_{21}t_1 + a_{22}t_2 + \ldots + a_{2m}t_m \\ \vdots \\ x_n = a_{n1}t_1 + a_{n2}t_2 + \ldots + a_{nm}t_m \end{matrix} \text{ for some } t_1, \ldots, t_m \in K \right\}.$$

For example, the set of all vectors (x, y, z) parameterized by the equations

$$x = 2t_1 + 3t_2, \quad y = 5t_1 - 4t_2, \quad \text{and} \quad z = -t_1 + 2t_2$$

is a two-dimensional subspace of K^3, if K is a number field (such as real or rational numbers).

Span of Vectors

In linear algebra, the system of parametric equations can be written as a single vector equation:

$$\begin{bmatrix} x \\ y \\ z \end{bmatrix} = t_1 \begin{bmatrix} 2 \\ 5 \\ -1 \end{bmatrix} + t_2 \begin{bmatrix} 3 \\ -4 \\ 2 \end{bmatrix}.$$

The expression on the right is called a linear combination of the vectors (2, 5, −1) and (3, −4, 2). These two vectors are said to span the resulting subspace.

In general, a linear combination of vectors $v_1, v_2, \ldots, v_k$ is any vector of the form

$$t_1 v_1 + \cdots + t_k v_k.$$

The set of all possible linear combinations is called the span:

$$\text{Span}\{v_1, \ldots, v_k\} = \{t_1 v_1 + \cdots + t_k v_k : t_1, \ldots, t_k \in K\}.$$

If the vectors $v_1, \ldots, v_k$ have n components, then their span is a subspace of K^n. Geometrically, the span is the flat through the origin in n-dimensional space determined by the points $v_1, \ldots, v_k$.

Example

The *xz*-plane in R^3 can be parameterized by the equations

$$x = t_1, \quad y = 0, \quad z = t_2.$$

As a subspace, the *xz*-plane is spanned by the vectors (1, 0, 0) and (0, 0, 1). Every vector in the *xz*-plane can be written as a linear combination of these two:

$$(t_1, 0, t_2) = t_1(1, 0, 0) + t_2(0, 0, 1).$$

Geometrically, this corresponds to the fact that every point on the *xz*-plane can be reached from the origin by first moving some distance in the direction of (1, 0, 0) and then moving some distance in the direction of (0, 0, 1).

Column Space and Row Space

A system of linear parametric equations in a finite-dimensional space can also be written as a single matrix equation:

$$x = At \quad \text{where} \quad A = \begin{bmatrix} 2 & 3 \\ 5 & -4 \\ -1 & 2 \end{bmatrix}.$$

In this case, the subspace consists of all possible values of the vector x. In linear algebra, this subspace is known as the column space (or image) of the matrix A. It is precisely the subspace of K^n spanned by the column vectors of A.

The row space of a matrix is the subspace spanned by its row vectors. The row space is interesting because it is the orthogonal complement of the null space.

Independence, Basis, and Dimension

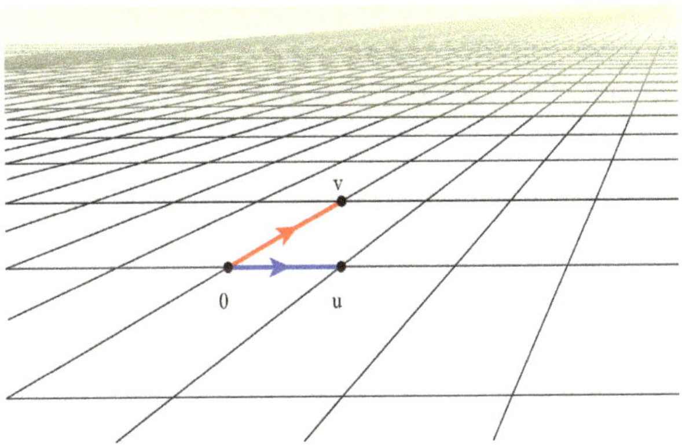

The vectors **u** and **v** are a basis for this two-dimensional subspace of R^3.

In general, a subspace of K^n determined by k parameters (or spanned by k vectors) has dimension k. However, there are exceptions to this rule. For example, the subspace of K^3 spanned by the three vectors (1, 0, 0), (0, 0, 1), and (2, 0, 3) is just the xz-plane, with each point on the plane described by infinitely many different values of t_1, t_2, t_3.

In general, vectors $v_1, \ldots, v_k$ are called linearly independent if

$$t_1 v_1 + \cdots + t_k v_k \neq u_1 v_1 + \cdots + u_k v_k$$

for $(t_1, t_2, \ldots, t_k) \neq (u_1, u_2, \ldots, u_k)$. If $v_1, \ldots, v_k$ are linearly independent, then the coordinates $t_1, \ldots, t_k$ for a vector in the span are uniquely determined.

A basis for a subspace S is a set of linearly independent vectors whose span is S. The number of elements in a basis is always equal to the geometric dimension of the subspace. Any spanning set for a subspace can be changed into a basis by removing redundant vectors.

Example

Let S be the subspace of R^4 defined by the equations

$$x_1 = 2x_2 \quad \text{and} \quad x_3 = 5x_4.$$

Then the vectors (2, 1, 0, 0) and (0, 0, 5, 1) are a basis for S. In particular, every vector that satisfies the above equations can be written uniquely as a linear combination of the two basis vectors:

$$(2t_1, t_1, 5t_2, t_2) = t_1(2,1,0,0) + t_2(0,0,5,1).$$

The subspace *S* is two-dimensional. Geometrically, it is the plane in R^4 passing through the points (0, 0, 0, 0), (2, 1, 0, 0), and (0, 0, 5, 1).

Operations and Relations on Subspaces

Inclusion

The set-theoretical inclusion binary relation specifies a partial order on the set of all subspaces (of any dimension).

A subspace cannot lie in any subspace of lesser dimension. If dim $U = k$, a finite number, and $U \subset W$, then dim $W = k$ if and only if $U = W$.

Intersection

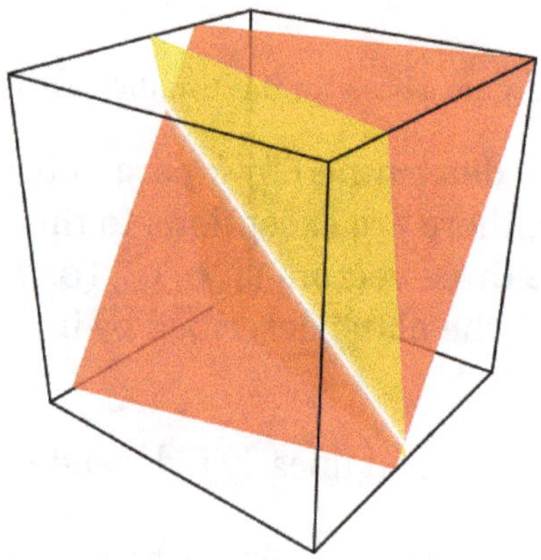

In R^3, the intersection of two-dimensional subspaces is one-dimensional

Given subspaces *U* and *W* of a vector space *V*, then their intersection $U \cap W := \{v \in V :$ v is an element of both *U* and *W*$\}$ is also a subspace of *V*.

Proof:

1. Let v and w be elements of $U \cap W$. Then v and w belong to both *U* and *W*. Because *U* is a subspace, then v + w belongs to *U*. Similarly, since *W* is a subspace, then v + w belongs to *W*. Thus, v + w belongs to $U \cap W$.

2. Let v belong to $U \cap W$, and let *c* be a scalar. Then v belongs to both *U* and *W*. Since *U* and *W* are subspaces, *c*v belongs to both *U* and *W*.

3. Since *U* and *W* are vector spaces, then 0 belongs to both sets. Thus, 0 belongs to $U \cap W$.

For every vector space *V*, the set {0} and *V* itself are subspaces of *V*.

Sum

If U and W are subspaces, their sum is the subspace

$$\mathrm{U}+\mathrm{W}=\{\mathrm{u}+\mathrm{w}:\mathrm{u}\in\mathrm{U},\mathrm{w}\in\mathrm{W}\}.$$

For example, the sum of two lines is the plane that contains them both. The dimension of the sum satisfies the inequality

$$\max(\dim U,\dim W)\leq\dim(U+W)\leq\dim(U)+\dim(W).$$

Here the minimum only occurs if one subspace is contained in the other, while the maximum is the most general case. The dimension of the intersection and the sum are related:

$$\dim(U+W)=\dim(U)+\dim(W)-\dim(U\cap W).$$

Lattice of Subspaces

The operations intersection and sum make the set of all subspaces a bounded modular lattice, where the {0} subspace, the least element, is an identity element of the sum operation, and the identical subspace V, the greatest element, is an identity element of the intersection operation.

Other

If V is an inner product space, then the orthogonal complement $\perp$ of any subspace of V is again a subspace. This operation, understood as negation ($\neg$), makes the lattice of subspaces a (possibly infinite) orthocomplemented lattice (it is not a distributive lattice).

In a pseudo-Euclidean space there are orthogonal complements too, but such operation does not form a Boolean algebra (nor a Heyting algebra) because of null subspaces, for which $N\cap N^{\perp}=N\neq\{0\}$. The same case presents the $^{\perp}$ operation in symplectic vector spaces.

Algorithms

Most algorithms for dealing with subspaces involve row reduction. This is the process of applying elementary row operations to a matrix until it reaches either row echelon form or reduced row echelon form. Row reduction has the following important properties:

1. The reduced matrix has the same null space as the original.
2. Row reduction does not change the span of the row vectors, i.e. the reduced matrix has the same row space as the original.
3. Row reduction does not affect the linear dependence of the column vectors.

Basis for a Row Space

Input An m × n matrix A.

Output A basis for the row space of A.

1. Use elementary row operations to put A into row echelon form.
2. The nonzero rows of the echelon form are a basis for the row space of A.

If we instead put the matrix A into reduced row echelon form, then the resulting basis for the row space is uniquely determined. This provides an algorithm for checking whether two row spaces are equal and, by extension, whether two subspaces of K^n are equal.

Subspace Membership

Input A basis $\{b_1, b_2, ..., b_k\}$ for a subspace S of K^n, and a vector v with n components.

Output Determines whether v is an element of S

1. Create a $(k + 1) \times n$ matrix A whose rows are the vectors $b_1, \dots, b_k$ and v.
2. Use elementary row operations to put A into row echelon form.
3. If the echelon form has a row of zeroes, then the vectors $\{b_1, ..., b_k, v\}$ are linearly dependent, and therefore $v \in S$.

Basis for a Column space

Input An $m \times n$ matrix A

Output A basis for the column space of A

1. Use elementary row operations to put A into row echelon form.
2. Determine which columns of the echelon form have pivots. The corresponding columns of the original matrix are a basis for the column space.

This produces a basis for the column space that is a subset of the original column vectors. It works because the columns with pivots are a basis for the column space of the echelon form, and row reduction does not change the linear dependence relationships between the columns.

Coordinates for a Vector

Input A basis $\{b_1, b_2, ..., b_k\}$ for a subspace S of K^n, and a vector $v \in S$

Output Numbers $t_1, t_2, ..., t_k$ such that $v = t_1 b_1 + \cdots + t_k b_k$

1. Create an augmented matrix A whose columns are $b_1,...,b_k$, with the last column being v.
2. Use elementary row operations to put A into reduced row echelon form.
3. Express the final column of the reduced echelon form as a linear combination of the first k columns. The coefficients used are the desired numbers $t_1, t_2, ..., t_k$. (These should be precisely the first k entries in the final column of the reduced echelon form.)

If the final column of the reduced row echelon form contains a pivot, then the input vector v does not lie in S.

Basis for a Null Space

Input An $m \times n$ matrix A.

Output A basis for the null space of A

1. Use elementary row operations to put A in reduced row echelon form.
2. Using the reduced row echelon form, determine which of the variables $x_1, x_2, ..., x_n$ are free. Write equations for the dependent variables in terms of the free variables.
3. For each free variable x_i, choose a vector in the null space for which $x_i = 1$ and the remaining free variables are zero. The resulting collection of vectors is a basis for the null space of A.

Basis for the Sum and Intersection of Two Subspaces

Given two subspaces U and W of V, a basis of the sum $\mathrm{U}+\mathrm{W}$ and the intersection $\mathrm{U}\cap\mathrm{W}$ can be calculated using the Zassenhaus algorithm

Equations for a Subspace

Input A basis $\{b_1, b_2, ..., b_k\}$ for a subspace S of K^n

Output An $(n - k) \times n$ matrix whose null space is S.

1. Create a matrix A whose rows are $b_1, b_2, ..., b_k$.
2. Use elementary row operations to put A into reduced row echelon form.
3. Let $c_1, c_2, ..., c_n$ be the columns of the reduced row echelon form. For each column without a pivot, write an equation expressing the column as a linear combination of the columns with pivots.
4. This results in a homogeneous system of $n - k$ linear equations involving the variables $c_1,...,c_n$. The $(n - k) \times n$ matrix corresponding to this system is the desired matrix with nullspace S.

Example

If the reduced row echelon form of A is

$$\begin{bmatrix} 1 & 0 & -3 & 0 & 2 & 0 \\ 0 & 1 & 5 & 0 & -1 & 4 \\ 0 & 0 & 0 & 1 & 7 & -9 \\ 0 & 0 & 0 & 0 & 0 & 0 \end{bmatrix}$$

then the column vectors $c_1, \ldots, c_6$ satisfy the equations

$$\begin{aligned} c_3 &= -3c_1 + 5c_2 \\ c_5 &= 2c_1 - c_2 + 7c_4 \\ c_6 &= 4c_2 - 9c_4 \end{aligned}$$

It follows that the row vectors of A satisfy the equations

$$\begin{aligned} x_3 &= -3x_1 + 5x_2 \\ x_5 &= 2x_1 - x_2 + 7x_4 \\ x_6 &= 4x_2 - 9x_4. \end{aligned}$$

In particular, the row vectors of A are a basis for the null space of the corresponding matrix.

Pseudo-Euclidean Space

In mathematics and theoretical physics, a pseudo-Euclidean space is a finite-dimensional real n-space together with a non-degenerate quadratic form q. Such a quadratic form can, given a suitable choice of basis $(e_1, \ldots, e_n)$, be applied to a vector $x = x_1e_1 + \ldots + x_ne_n$, giving

$q(x) = \left(x_1^2 + \cdots + x_k^2\right) - \left(x_{k+1}^2 + \cdots + x_n^2\right)$ which is called the *magnitude* of the vector x.

For Euclidean spaces, $k = n$, implying that the quadratic form is positive-definite. When $0 \neq k \neq n$, q is an isotropic quadratic form. Note that if $i \leq k$ and $j > k$, then $q(e_i + e_j) = 0$, so that $e_i + e_j$ is a null vector. In a pseudo-Euclidean space with $k \neq n$, unlike in a Euclidean space, there exist vectors with negative magnitude.

As with the term *Euclidean space*, *pseudo-Euclidean space* may refer to either anaffine space or a vector space, though the latter may also be referred to as a pseudo-Euclidean vector space.

Geometry

The geometry of a pseudo-Euclidean space is consistent in spite of a breakdown of the some properties of Euclidean space; most notably that it is not a metric space as explained below. The affine structure is unchanged, and thus also the concepts line, plane and, generally, of an affine subspace (flat), as well as line segments.

Positive, Zero, and Negative Magnitudes

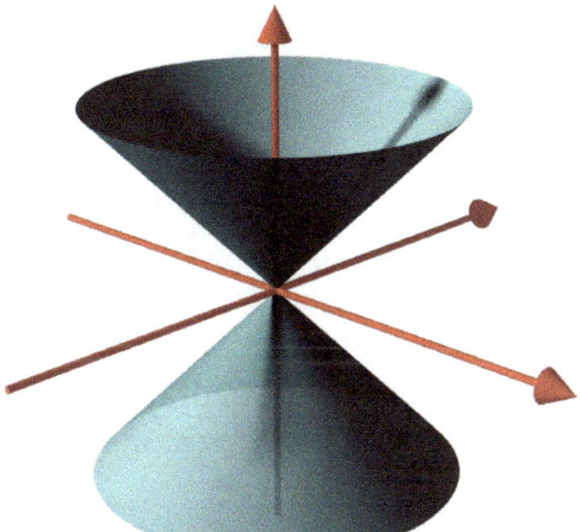

$n = 3$, k is either 1 or 2 depending on the choice of sign of q

A null vector is a vector for which the quadratic form is zero. Unlike in a Euclidean space, the vector can be non-zero, in which case it is orthogonal to itself. If the quadratic form is indefinite, a pseudo-Euclidean space has a linear cone of null vectors given by $\{ x : q(x) = 0 \}$. When the pseudo-Euclidean space provides a model for spacetime, the null cone is called the light cone of the origin.

The null cone separates two open sets, for which $q(x) > 0$ and $q(x) < 0$. If $k \geq 2$, then the set of vectors for which $q(x) > 0$ is connected. If $k = 1$, then it consists of two disjoint parts, one with $x_1 > 0$ and another with $x_1 < 0$. Similar statements can be made for vectors for which $q(x) < 0$ if k is replaced with $n - k$.

Interval

The magnitude q corresponds to the square of a vector (or its norm) in Euclidean case. To define the vector norm (and distance) in an invariant manner, one has to get square roots of magnitudes, which leads to possibly imaginary distances; see square root of negative numbers. But even for a triangle with positive magnitudes of all three sides (whose square roots are real and positive), the triangle inequality is not necessarily true.

Hence terms *norm* and *distance* are avoided in pseudo-Euclidean geometry, replaced with *magnitude* and *interval* respectively.

Though, for a curve whose tangent vectors all have the same sign of magnitude, the arc length is defined. It has important applications.

Rotations and Spheres

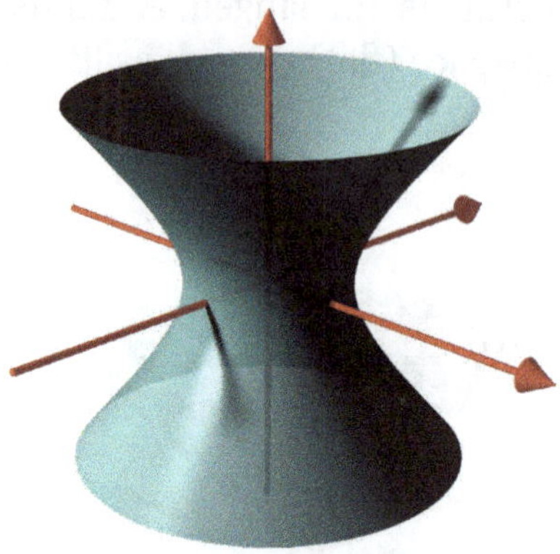

The rotations group of such space is indefinite orthogonal group $O(q)$, also denoted as $O(k, n-k)$ without a reference to particular quadratic form. Such "rotations" preserve the form q and, hence, the magnitude of each vector including whether it is positive, zero, or negative.

Whereas Euclidean space has a unit sphere, pseudo-Euclidean space has the hypersurfaces $\{x : q(x) = 1\}$ and $\{x : q(x) = -1\}$. Such a hypersurface, called a quasi-sphere, is preserved by the appropriate indefinite orthogonal group.

Symmetric Bilinear Form

The quadratic form q gives rise to a symmetric bilinear form defined as follows:

$$\langle x, y\rangle = \frac{1}{2}[q(x+y) - q(x) - q(y)] = (x_1 y_1 + \cdots + x_k y_k) - (x_{k+1} y_{k+1} + \cdots + x_n y_n).$$

The quadratic form can be expressed in terms of the bilinear form: $q(x) = \langle x, x\rangle$.

When $\langle x, y\rangle = 0$, then x and y are orthogonal vectors of the pseudo-Euclidean space. This bilinear form is often referred to as the scalar product, and sometimes as "inner product" or "dot product", but it does not define an inner product space and it does not have the properties of the dot product of Euclidean vectors.

If x and y are orthogonal and $q(x)q(y) < 0$, then x is hyperbolic-orthogonal to y.

The standard basis of the real n-space is orthogonal. There are no ortho*normal* bases in a pseudo-Euclidean space for which the bilinear form is indefinite, because it cannot be used to define a vector norm.

Subspaces and Orthogonality

For a (positive-dimensional) subspace U of a pseudo-Euclidean space, when the magnitude form q is restricted to U, following three cases are possible:

1. $q|_U$ is either positive or negative definite. Then, U is essentially Euclidean (up to the sign of q).

2. $q|_U$ is indefinite, but non-degenerate. Then, U is itself pseudo-Euclidean. It is possible only if $\dim U \geq 2$; if $\dim U = 2$, which means than U is a plane, then it is called a hyperbolic plane.

3. $q|_U$ is degenerate.

One of most jarring properties (for a Euclidean intuition) of pseudo-Euclidean vectors and flats is their orthogonality. When two non-zero Euclidean vectors are perpendicular, they are not collinear. Any Euclidean linear subspace intersects with its orthogonal complement only by the {0} subspace. But the definition immediately implies that any vector v of zero magnitude is perpendicular to itself. Hence, for the 1-subspace $N = \langle v \rangle$ generated by such non-zero vector, its orthogonal complement $N^\perp$ will be a superspace of N.

The formal definition of the orthogonal complement of a vector subspace in a pseudo-Euclidean space gives a perfectly well-defined result, which satisfies the equality $\dim U + \dim U^\perp = n$ due to the magnitude form's non-degeneracy. It is just the condition

$U \cap U^\perp = \{0\}$ or, equivalently, $U + U^\perp =$all space,

which can be broken if the subspace U contains a null direction. While subspaces form a lattice, as in any vector space, this $\perp$ operation is not an orthocomplementation, in contrast to inner product spaces.

For a subspace N composed *entirely* of null vectors (which means that the magnitude q, restricted to N, equals to 0), always holds:

$N \subset N^\perp$ or, equivalently, $N \cap N^\perp = N$.

Such a subspace can have up to $\min(k, n - k)$ dimensions.

For a (positive) Euclidean k-subspace its orthogonal complement is a $(n - k)$-dimensional negative "Euclidean" subspace, and vice versa. Generally, for a $(d_+ + d_- + d_0)$-dimensional subspace U consisting of d_+ positive and d_- negative dimensions, its orthogonal "complement" $U^\perp$ has $(k - d_+ - d_0)$ positive and $(n - k - d_- - d_0)$ negative dimensions, while the rest d_0 ones are degenerate and form the $U \cap U^\perp$ intersection.

Parallelogram Law and Pythagorean Theorem

The parallelogram law takes the form

$$q(x) + q(y) = \frac{1}{2}(q(x+y) + q(x-y)).$$

Using the square of the sum identity, for an arbitrary triangle one can express the magnitude of the third side from magnitudes of two sides and their bilinear form product:

$$q(x+y) = q(x) + q(y) + 2\langle x, y\rangle.$$

This demonstrates that, for orthogonal vectors, a pseudo-Euclidean analog of the Pythagorean theorem holds:

$$\langle x, y\rangle = 0 \Rightarrow q(x) + q(y) = q(x+y).$$

Angle

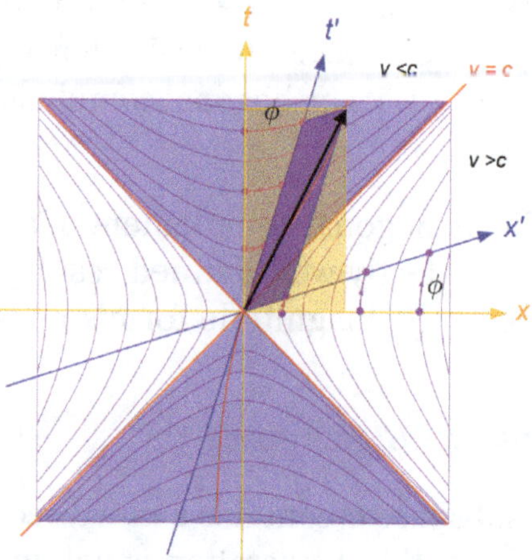

Generally, absolute value $|\langle x,y\rangle|$ of the bilinear form on two vectors may be greater than $\sqrt{|q(x)q(y)|}$, equal to it, or less. This causes similar problems with definition of angle as appeared above for distances. If $k = 1$ (only one positive term in q), then for positive-magnitude vectors:

$$|\langle x, y\rangle| \geq \sqrt{q(x)q(y)},$$

which permits to define hyperbolic angle, an analog of angle between these vectors through inverse hyperbolic cosine:

$$\operatorname{arcosh}\frac{|\langle x, y\rangle|}{\sqrt{q(x)q(y)}}.$$

It corresponds to the distance on a $(n - 1)$-dimensional hyperbolic space. This is known as rapidity in the context of theory of relativity discussed below. Unlike Euclidean angle, it takes values from $[0, +\infty)$ and equals to 0 for antiparallel vectors.

There is no reasonable definition of the angle between a null vector and another vector (either null or non-null).

Algebra and Tensor Calculus

Like Euclidean spaces, every pseudo-Euclidean space possesses a geometric algebra. Unlike properties above, where replacement of q to $-q$ changed numbers but not geometry, the sign reversal of the magnitude form actually *alters* Cℓ, so for example $C\ell_{1,2}(R)$ and $C\ell_{2,1}(R)$ are not isomorphic.

Just like over any vector space, there are pseudo-Euclidean tensors. Like with a Euclidean structure, there are raising and lowering indices operators but, unlike the case with Euclidean tensors, there is no bases where these operations do not change values of components. If there is a vector v^β, the corresponding covariant vector is:

$$v_\alpha = q_{\alpha\beta} v^\beta,$$

and with the standard-form

$$q_{\alpha\beta} = \begin{pmatrix} I_{k\times k} & 0 \\ 0 & -I_{(n-k)\times(n-k)} \end{pmatrix}$$

the first k components of v_α are numerically the same as ones of v^β, but the rest $n - k$ have opposite signs.

The correspondence between contravariant and covariant tensors makes a tensor calculus on pseudo-Riemannian manifolds analogous to one on Riemannian manifolds.

Examples

A very important pseudo-Euclidean space is Minkowski space, which is the mathematical setting in which Albert Einstein's theory of special relativity is formulated. For Minkowski space, $n = 4$ and $k = 3$ so that

$$q(x) = x_1^2 + x_2^2 + x_3^2 - x_4^2,$$

The geometry associated with this pseudo-metric was investigated by Poincaré. Its rotation group is the Lorentz group. The Poincaré group includes also translations and plays the same role as Euclidean groups of ordinary Euclidean spaces.

Another pseudo-Euclidean space is the plane $z = x + yj$ consisting of split-complex numbers, equipped with the quadratic form

$$\| z \| = zz^* = z^* z = x^2 - y^2.$$

This is the simplest case of an idefinite pseudo-Euclidean space ($n = 2$, $k = 1$) and the only one where the null cone dissects the space to *four* open sets. The group $SO^+(1,1)$ consists of so named hyperbolic rotations.

References

- Lang, Serge (2002), Algebra, Graduate Texts in Mathematics, 211 (Revised third ed.), New York: Springer-Verlag, ISBN 978-0-387-95385-4, MR 1878556, Zbl 0984.00001
- John B. Fraleigh, Victor J. Katz. A first course in abstract algebra. Addison-Wesley Publishing Company. 5 ed., 1967. ISBN 0-201-53467-3
- Novikov, S. P.; Fomenko, A.T.; [translated from the Russian by M. Tsaplina] (1990). Basic elements of differential geometry and topology. Dordrecht; Boston: Kluwer Academic Publishers. ISBN 0-7923-1009-8
- Szekeres, Peter (2004). A course in modern mathematical physics: groups, Hilbert space, and differential geometry. Cambridge University Press. ISBN 0-521-82960-7
- Blass, Andreas (1984), "Existence of bases implies the axiom of choice", Axiomatic set theory, Contemporary Mathematics volume 31, Providence, R.I.: American Mathematical Society, pp. 31–33, ISBN 0-8218-5026-1, MR 763890

2

Inner Product Space: An Additional Structure of Vector Space

Inner product space is a vector space along with a function called inner product. An inner product is also a normed vector space as it produces an associated norm. The topics discussed in the chapter are of great importance to broaden the existing knowledge on vector space.

Inner Product Space

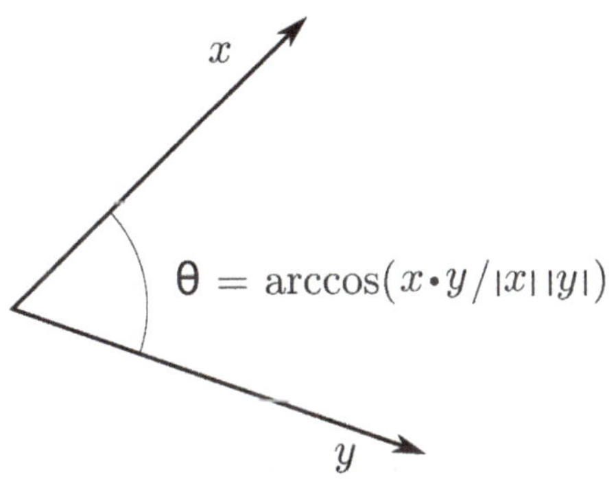

Geometric interpretation of the angle between two vectors defined using an inner product

In linear algebra, an inner product space is a vector space with an additional structure called an inner product. This additional structure associates each pair of vectors in the space with a scalar quantity known as the inner product of the vectors. Inner products allow the rigorous introduction of intuitive geometrical notions such as the length of a vector or the angle between two vectors. They also provide the means of defining orthogonality between vectors (zero inner product). Inner product spaces generalize Euclidean spaces (in which the inner product is the dot product, also known as the scalar product) to vector spaces of any (possibly infinite) dimension, and are studied in functional analysis.

An inner product naturally induces an associated norm, thus an inner product space is also a normed vector space. A complete space with an inner product is called a Hilbert space. An (incomplete) space with an inner product is called a pre-Hilbert space, since its completion with respect to the norm induced by the inner product is a Hilbert space. Inner product spaces over the field of complex numbers are sometimes referred to as unitary spaces.

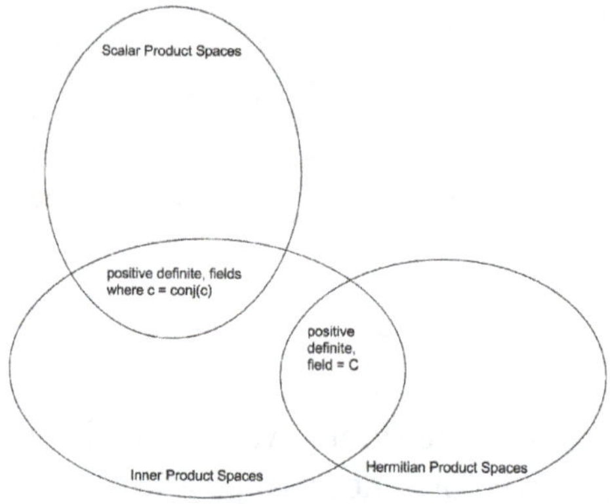

Scalar Product Spaces, over any field, have "scalar products" that are symmetrical and linear in the first (by convention) argument. Hermitian Product Spaces are restricted to the field of complex numbers and have "hermitian products" that are conjugate-symmetrical and linear in the first argument (by convention). Inner Product Spaces may be defined over any field, having "inner products" that are linear in the first (by convention) argument, conjugate-symmetrical, and positive-definite. Unlike Inner Products, Scalar Products and Hermitian Products need not be positive-definite.

Definition

In this article, the field of scalars denoted F is either the field of real numbers R or the field of complex numbers C.

Formally, an inner product space is a vector space *V* over the field *F* together with an *inner product*, i.e., with a map

$$\langle \cdot,\cdot \rangle : \mathrm{V} \times \mathrm{V} \rightarrow \mathrm{F}$$

that satisfies the following three axioms for all vectors $\mathrm{x}, \mathrm{y}, \mathrm{z} \in \mathrm{V}$ and all scalars $a \in F$:

- Conjugate symmetry:

$$\langle \mathrm{x}, \mathrm{y} \rangle = \overline{\langle \mathrm{y}, \mathrm{x} \rangle}$$

- Linearity in the first argument:

$$\langle \mathrm{ax}, \mathrm{y} \rangle = \mathrm{a}\langle \mathrm{x}, \mathrm{y} \rangle$$

$$\langle \mathrm{x} + \mathrm{y}, \mathrm{z} \rangle = \langle \mathrm{x}, \mathrm{z} \rangle + \langle \mathrm{y}, \mathrm{z} \rangle$$

- Positive-definiteness:

$$\langle \mathrm{x}, \mathrm{x} \rangle \geq 0$$

$$\langle \mathrm{x}, \mathrm{x} \rangle = 0 \Leftrightarrow \mathrm{x} = 0 .$$

Alternative Definitions, Notations and Remarks

Some authors, especially in physics and matrix algebra, prefer to define the inner product and the sesquilinear form with linearity in the second argument rather than the first. Then the first argument becomes conjugate linear, rather than the second. In those disciplines we would write the product $\langle x, y \rangle$ as $\langle y | x \rangle$ (the bra–ket notation of quantum mechanics), respectively $y^{\dagger}x$ (dot product as a case of the convention of forming the matrix product *AB* as the dot products of rows of *A* with columns of *B*). Here the kets and columns are identified with the vectors of *V* and the bras and rows with the dual vectors or linear functionals of the dual space V^*, with conjugacy associated with duality. This reverse order is now occasionally followed in the more abstract literature, taking $\langle x, y \rangle$ to be conjugate linear in *x* rather than *y*. A few instead find a middle ground by recognizing both $\langle \cdot, \cdot \rangle$ and $\langle \cdot | \cdot \rangle$ as distinct notations differing only in which argument is conjugate linear.

There are various technical reasons why it is necessary to restrict the basefield to R and **C** in the definition. Briefly, the basefield has to contain an ordered subfield in order for non-negativity to make sense, and therefore has to have characteristic equal to 0 (since any ordered field has to have such characteristic). This immediately excludes finite fields. The basefield has to have additional structure, such as a distinguished automorphism. More generally any quadratically closed subfield of R or C will suffice for this purpose, e.g., the algebraic numbers or the constructible numbers. However in these cases when it is a proper subfield (i.e., neither R nor C) even finite-dimensional inner product spaces will fail to be metrically complete. In contrast all finite-dimensional inner product spaces over R or C, such as those used in quantum computation, are automatically metrically complete and hence Hilbert spaces.

In some cases we need to consider non-negative *semi-definite* sesquilinear forms. This means that $\langle x, x \rangle$ is only required to be non-negative. We show how to treat these below.

Elementary Properties

When *F* = R, conjugate symmetry reduces to symmetry. That is, $\langle x, y \rangle = \langle y, x \rangle$ for *F* = R; while for F = C, $\langle x, y \rangle$ is equal to the complex conjugate.

Notice that conjugate symmetry implies that $\langle x, x \rangle$ is real for all *x*, since we have:

$$\langle x, x \rangle = \overline{\langle x, x \rangle}.$$

Moreover, sesquilinearity implies that

$$\langle -x, x \rangle = -1\langle x, x \rangle = \overline{-1}\langle x, x \rangle = \langle x, -x \rangle.$$

Conjugate symmetry and linearity in the first variable gives

$$\langle x, ay\rangle = \overline{\langle ay, x\rangle} = \bar{a}\overline{\langle y, x\rangle} = \bar{a}\langle x, y\rangle$$

$$\langle x, y+z\rangle = \overline{\langle y+z, x\rangle} = \overline{\langle y, x\rangle} + \overline{\langle z, x\rangle} = \langle x, y\rangle + \langle x, z\rangle,$$

so an inner product is a sesquilinear form. Conjugate symmetry is also called Hermitian symmetry, and a conjugate symmetric sesquilinear form is called a *Hermitian form*. While the above axioms are more mathematically economical, a compact verbal definition of an inner product is a *positive-definite Hermitian form.*

In the case of F = R, conjugate-symmetry reduces to symmetry, and sesquilinear reduces to bilinear. So, an inner product on a real vector space is a *positive-definite symmetric bilinear form.*

From the linearity property it is derived that x = o implies $\langle x, x\rangle = 0$, while from the positive-definiteness axiom we obtain the converse, $\langle x, x\rangle = 0$, implies x = o. Combining these two, we have the property that $\langle x, x\rangle = 0$, if and only if x = o.

Combining the linearity of the inner product in its first argument and the conjugate symmetry gives the following important generalization of the familiar square expansion:

$$\langle x+y, x+y\rangle = \langle x, x\rangle + \langle x, y\rangle + \langle y, x\rangle + \langle y, y\rangle.$$

Assuming the underlying field to be R, the inner product becomes symmetric, and we obtain

$$\langle x \pm y, x \pm y\rangle = \langle x, x\rangle \pm 2\langle x, y\rangle + \langle y, y\rangle,$$

The property of an inner product space V that

$$\langle x+y, z\rangle = \langle x, z\rangle + \langle y, z\rangle,$$

$$\langle x, y+z\rangle = \langle x, y\rangle + \langle x, z\rangle$$

is also known as *additivity*.

Examples

- A simple example is the real numbers with the standard multiplication as the inner product

  $$\langle x, y\rangle := xy.$$

  More generally, the real n-space R^n with the dot product is an inner product space, an example of a Euclidean n-space.

$$\left\langle \begin{bmatrix} x_1 \\ \vdots \\ x_n \end{bmatrix}, \begin{bmatrix} y_1 \\ \vdots \\ y_n \end{bmatrix} \right\rangle := x^T y = \sum_{i=1}^{n} x_i y_i = x_1 y_1 + \cdots + x_n y_n,$$

where x^T is the transpose of x.

- The general form of an inner product on $\mathbf{C}^n$ is known as the Hermitian form and is given by

$$\langle x, y \rangle := y^{\dagger} M x = \overline{x^{\dagger} M y},$$

where M is any Hermitian positive-definite matrix and $\mathbf{y}^{\dagger}$ is the conjugate transpose of $\mathbf{y}$. For the real case this corresponds to the dot product of the results of directionally different scaling of the two vectors, with positive scale factors and orthogonal directions of scaling. Up to an orthogonal transformation it is a weighted-sum version of the dot product, with positive weights.

- The article on Hilbert space has several examples of inner product spaces wherein the metric induced by the inner product yields a complete metric space. An example of an inner product which induces an incomplete metric occurs with the space $C([a, b])$ of continuous complex valued functions on the interval $[a, b]$. The inner product is

$$\langle f, g \rangle := \int_a^b f(t)\overline{g(t)}dt.$$

This space is not complete; consider for example, for the interval [−1,1] the sequence of continuous "step" functions, $\{f_k\}_k$, defined by:

$$f_k(t) = \begin{cases} 0 & t \in [-1, 0] \\ 1 & t \in \left[\frac{1}{k}, 1\right] \\ kt & t \in \left(0, \frac{1}{k}\right) \end{cases}$$

This sequence is a Cauchy sequence for the norm induced by the preceding inner product, which does not converge to a *continuous* function.

- For real random variables X and Y, the expected value of their product

$$\langle X, Y \rangle := E(XY)$$

is an inner product. In this case, $\langle X, X \rangle = 0$ if and only if $\Pr(X = 0) = 1$ (i.e., $X = 0$ almost surely). This definition of expectation as inner product can be extended to random vectors as well.

- For real matrices of the same size, $\langle A,B\rangle := \mathrm{tr}(AB^T)$ with transpose as conjugation

$$\left(\langle A,B\rangle = \langle B^T, A^T\rangle\right)$$

is an inner product.

Recall that X is called a vector space over a field $\mathbb{F}$ when X is an abelian group equipped with an $X\times\mathbb{F}\xrightarrow{(x,y)\to x\lambda}X$ satisfying expected identities like $(x + u)\lambda = x\lambda + u\lambda$, $x(\lambda + \mu) = x\lambda + x\mu$ etc; we also can, and do, write $x\lambda \in X$ as λx (this is not a problem since F is commutative).

Further, given two vector spaces X and Y (over the same field $\mathbb{F}$) a linear transformation $X\xrightarrow{A}Y$ is a function such that $A(x + u\lambda) = A(x) + A(u)\lambda$ for all $x, u \in X, \lambda \in \mathbb{F}$. For $X\xrightarrow{A,B}Y$ we define $\lambda A + B$ by $(\lambda A + B)(x) = \lambda A(x) + B(x)$ at each $x \in X$. Then the collection L(X, Y) of all linear transformations $X \to Y$ becomes a vector space. Verify this fact. What is $o \in L(X, Y)$? Is $X\xrightarrow{idX}X$ in L(X, X)?

A linear transformation $X\xrightarrow{A}Y$ is called an isomorphism iff it is a bijection (prove that then $Y\xrightarrow{A^{-1}}X$ is also and further, $A^{-1}A = id_X$, $AA^{-1} = id_Y$). If there is an isomorphism $X\xrightarrow{A}Y$ we write $X\cong Y$ and regard the two vector spaces X and Y as 'the same', identifying $x \in X$ with $A(x) \in Y$.

Prove that $L(\mathbb{F},X)\cong X$.

Hint:

Define $L(\mathbb{F},X)\xrightarrow{f}X$ by $f(A) := A(1)$. Then $f(\lambda A + B) = (\lambda A + B)(1) = \lambda A(1) + B(1) = \lambda f(A) + f(B)$ so that f is a linear transformation.

Next, for $x \in X$, define $X\xrightarrow{g}L(\mathbb{F},X)$ by writing $(g(x))(\mu) := \mu x$. Then $(g(x))(\lambda\mu + \nu) = (\lambda\mu + \nu)(x) = \lambda(\mu x) + \nu x = \lambda g(x)(\mu) + g(x)(\nu)$ so that g(x) is indeed a linear transformation $\mathbb{F}\xrightarrow{X}$ and we do have $X\xrightarrow{g}L(\mathbb{F},X)$. Further, we have $(g(\lambda x + u))(\mu) = \mu(\lambda x) + \nu x = \lambda\mu x + \mu u = (\lambda g(x) + g(u))(\mu)$ at each $\mu \in \mathbb{F}$ so that the two functions $g(\lambda x + u)$ and $\lambda g(x) + g(u)$ are the same and we get $g(\lambda x + u) = \lambda g(x) + g(u)$. This means g is a linear transformation. Then $((gf)(A))(\lambda) = (g(A(1)))(\lambda) = \lambda A(1) = A(\lambda.1) = A(\lambda)$ at each $\lambda \in \mathbb{F}$ so that $gf(A) = A$ at each $A \in L(\mathbb{F}, X)$. Thus we have $gf = idL(\mathbb{F}, X)$. Also, $(fg(x)) = (g(x))(1) = 1.x = x$ at each $x \in X$ which means $fg = id_X$. This proves that $X\cong L(\mathbb{F},X)$ as desired.

Remark 0.1 In view of this, we can think of a vector $x \in X$ as a linear transformation $\mathbb{F}\xrightarrow{x}X$ and a scalar $\lambda \in \mathbb{F}$ can be regarded as a linear transformation $\mathbb{F}\xrightarrow{\lambda}\mathbb{F}$. Viewed like this, $x\lambda \in X$ is simply the composition $\mathbb{F}\xrightarrow{\lambda}\mathbb{F}\xrightarrow{x}X$. That this can be also thought as λx requires a more complicated argument which is beyond our scope.

The vector space L (X, $\mathbb{F}$) will be written as X' and called the transpose (sometimes, the dual) of X. An element $\varphi \in X'$ will be called a (linear) form (sometimes, functional) on X. We shall write $\varphi(x) \in \mathbb{F}$ as $<\varphi \mid x>$ (this is the Dirac notation). A function $X' \times X \xrightarrow{\beta} \mathbb{F}$ satisfying $\beta(\lambda\varphi + \alpha, x + u\mu) = \lambda\beta(\varphi, x) + \beta(\alpha, x) + \lambda(\varphi, u)\mu + \beta(\alpha, u)\mu$ for all $\varphi, \alpha \in X'$, $x, u \in X$, $\lambda, \mu \in \mathbb{F}$ (thus it is simply a linear form in each variable separately) will be called bilinear.

Show that $X' \times X \xrightarrow{<-\mid->} \mathbb{F}$ is bilinear.

Hint:

$<\lambda\varphi + \alpha \mid x + u\mu> = \lambda <\varphi \mid x> + <\alpha \mid x> + \lambda <\varphi \mid u> \mu + <\alpha \mid u> \mu$. (To see this, simply note that $(\lambda\varphi + \alpha)(x + u\mu)$ is the LHS and φ, α are linear forms on X).

For a linear transformation $X \xrightarrow{A} Y$ define $Y' \xrightarrow{A'} X'$ by writing $<A'(\psi)|x> := <\psi|A(x)>$ for each $\psi \in Y', x \in X$ and show that this is well defined linear transformation.

Hint:

$\psi = \varsigma$ means that at each $x \in X$ we have

$$(A'(\psi))(x) = <A'(\psi)|x> = <\psi|A(x)> = <\varsigma|A(x)> = <A'(\varsigma)|x> = (A'(\varsigma))(x).$$

Thus a single ψ does not have two candidates for $A'(\psi)$.

Since $<A'(\psi)|x + u\lambda> = <\psi|A(x + u\lambda)> = <\psi|Ax> + <\psi|Au> \lambda$ ($\because$ A is a linear tr) $= <A'(\psi)|x> + <A'(\psi)|u> \lambda$,

we conclude that $A'(\psi) \in X'$.

Since $<A'(\lambda\psi + \varsigma)|x> = <\lambda\psi + \varsigma|Ax> = \lambda <\psi|Ax> + <\varsigma|Ax> = A'(\psi)|x> + <A'(\varsigma)|x>$

we conclude that at each $x \in X$ we have $(A'(\lambda\psi + \varsigma))(x) = \lambda(A'(\psi))(x) + (A'(\varsigma)(x))$

so that $A'(\lambda\psi + \varsigma) = \lambda A'(\psi) + A'(\varsigma)$; this says A' is linear.

Write $<\varphi \mid x>$ as $<\Gamma(x) \mid \varphi>$ and extract a linear transformation $X \xrightarrow{\Gamma} X''$ from this.

Hint:

If $x = u$, we have $<\Gamma(x) \mid \varphi> = <\varphi \mid x> = <\varphi \mid u> = <\Gamma(u) \mid \varphi>$ at each $\varphi \in X'$ so that $\Gamma(x) = \Gamma(u)$ and we have no two candidates for $\Gamma(x)$, given a single $x \in X$. Since $<\Gamma(x + u\lambda) \mid \varphi> = <\varphi \mid (x + u\lambda)> = <\varphi \mid x> + <\varphi \mid u> \lambda = <\Gamma x \mid \varphi> + \Gamma(u)\lambda$, we have $X \xrightarrow{\Gamma} X''$ linear (since $\Gamma(x) \in X''$ is clear: $<\Gamma(x)|\lambda\varphi + \varsigma> = <\lambda\varphi + \varsigma|x> = \lambda <\varphi|x> + <\varsigma|x> = \lambda <\Gamma(x)|\varsigma>$

If $S = \{x_i \mid i \in I\}$ is any indexed collection of vectors in X, we say that $x = \sum_{i \in I} x_i \lambda_i$ is a linear combination of vectors from S; in this expression, it is understood that all but finitely may $\lambda_i \in \mathbb{F}$ are zero so that $x = \sum_{i \in I} x_i \lambda_i$ is a finite sum and thus we have x ∈ X. In particular, we agree that $0 = \sum_{i \in \phi} x_i \lambda_i$ i.e. the vacuous linear combination is the zero vector. Also, $\lambda_i = 0$ for all i ∈ I supplies the 'trivial combination' and yields the zero vector.

The set S is linearly dependent if the zero vector can be expressed as a non trivial linear combination of distinct elements from S; S is called linearly independent if it is not linearly dependent. In particular, the empty set is linearly independent. Clearly then, S is linearly independent iff there is at most one way to write x as $x = \sum_i x_i \lambda_i$ (indeed, $\sum x_i \lambda_i = \sum x_i \mu_i$ means $0 = \sum x_i (\lambda_i - \mu_i)$ which means $\lambda_i - \mu_i = 0$ for each i since S is linearly independent).

The collection $\left\{ x \in X \,\middle|\, \sum_{i \in I} x_i \lambda_i, x_i \in S \right\}$ is clearly a subspace of X and is called the subspace spanned by S; we shall denote this subspace by << S >> A collection $S = \{x_i \mid i \in I\}$ will be called a basis of X iff << S >>= X and S is linearly independent i.e. S is a basis of X iff each x ∈ X can be written in exactly one way as a linear combination $x = \sum x_i \lambda_i, x_i \in S, \lambda_i \in \mathbb{F}$ only finitely many λ_i being nonzero of course.

We shall supply proofs of the following in an appendix to this handout; here we accept them and feel free to employ them.

(a) Every vector space has a basis

(b) The number of elements in any basis of a vector space X is the same; this number is called the dimension of X and we write it as dim X. In particular, we say that X is finite-dimensional iff dim X < ∞. We agree that φ is a basis of the vector space {o} and thus there is exactly this space which has dimension o.

(c) If {Xi} is a collection of vector spaces, their product set $X = \prod X_i$ is turned into a vector space by writing $(x + y)_i := x_i + y_i$, $(\lambda x)_i = (x_i\lambda) = (x\lambda)_i = (\lambda x_i)$ (where by $x = (x_i)$ we mean the element x ∈ X whose i-th coordinate is $x_i \in X_i$) and the projections $X \xrightarrow{\text{proj}_i} X_i$ given by $\text{proj}_i\,(x) = x_i$ are linear. Further, their sum, the disjoint union, can be turned into a vector space $\oplus X_i$ by requiring the injections $X_i \xrightarrow{\text{inj}_i} \oplus X_i$ given by $\text{inj}_i\,(x_i) = x_i$ to be linear.

We are interested only in the finite version of these cases and shall not enter into the somewhat intricate definitions, arguments, and differentials involved. Suffice it to say that

(i) If P and Q are two subspaces of X, P ∩ Q is of course a subspace but P ∩ Q may not be; the subspace << P ∩ Q >> is written P ⊕ Q when P ∪ Q = {o} is known; this is the only case that interests us.

We say $P \oplus Q$ is the direct sum of P and Q; its elements are $p + q$ for uniquely given $p \in P$, $q \in Q$ and $P \cap Q = \{o\}$ is of course true. One thinks of $p \in P$ as $p + o \in P \oplus Q$ and $q \in Q$ as $o + q \in P \oplus Q$. Further, $P \oplus Q \to P$ given by $p + q \to p$ and $P \oplus Q \to Q$ given by $p + q \to q$ can be regarded as projections permitting us to think of $P \oplus Q$ as the product of P and Q while $p \to p + o$ and $q \to o + q$ supply injections $Q \to P \oplus Q$, $Q \to P \oplus Q$ permitting us to think of $P \oplus Q$ as the "sum" of P and Q.

(ii) What has been said for two subspaces P and Q in (i) above, remains valid for a finite sum $X_1 \oplus \cdots \oplus X_n$ i.e. each $x \in X_1 \oplus \cdots \oplus X_n$ can be written uniquely as $x = x_1 + \cdots + x_n$, $x_i \in X_i$, $X_i \cap X_j = \{o\}$ if $i \neq j$ and there are injections $X_i \to X_1 \oplus \ldots X_n$ given by $x_i \to o + \cdots + x_i + o + \ldots$ as well as projection given by $x_1 + \cdots + x_n \to x_i$, $1 \le i \le n$. When $X = X_1 \oplus \cdots \oplus X_n$, the RHS is said to be a decomposition of X into $X_1, \ldots, X_n$.

We do not include (c) in our appendix and this will be not discussed further in this course. But note the following to help understanding the ideals that have come up so far.

Example : Consider $X = \sum_{i=1}^{n} X_i$ each $X_i = \mathbb{F}$. Thus $X = \mathbb{F}^n := \mathbb{F} \oplus \cdots \oplus \mathbb{F}$ (n copies). Its elements will be written as column vectors

$$x = \begin{pmatrix} x^1 \\ \cdot \\ \cdot \\ \cdot \\ x^n \end{pmatrix} = \begin{pmatrix} x^1 \\ 0 \\ \cdot \\ \cdot \\ 0 \end{pmatrix} + \begin{pmatrix} 0 \\ x^2 \\ 0 \\ \cdot \\ 0 \end{pmatrix} + \cdots + \begin{pmatrix} 0 \\ \cdot \\ \cdot \\ \cdot \\ x^n \end{pmatrix} = \sum_{i=1}^{n} e_i x^i,\ x^i \in \mathbb{F},\ e_i = \begin{pmatrix} 0 \\ \cdot \\ 1 \\ 0 \\ 0 \end{pmatrix} \to \text{i-th row.}$$

This has then $e = \{e_i\}_{i=1}^{n}$ as a basis (prove this by showing that e is linearly independent and spans $\mathbb{F}^n$). If V has dimension $n < \infty$, choose a basis $\{b_1, \cdots, b_n\}$ for V (which must exist and have exactly n distinct elements by (a) and (b) above). Define $\mathbb{F}^n \xrightarrow{A} V$ by $A(e_i) = b_i$ so that for $x = \sum_{i=1}^{n} e_i x^i$ we have $Ax = \sum_{i=1}^{n} A(e_i) x^i$ (by linearity of A which is agreed upon) $= \sum_{i=1}^{n} b_i x^i$. Then since any $v \in V$ is $v = \sum_{i=1}^{n} b_i v^i$ for exactly one n-tuple $\{v^1, \cdots, v^n\}$ of scalars (= elements of $\mathbb{F}$), we have $A^{-1}(v) := \sum_{i=1}^{n} A^{-1}(b_i) v^i = \sum_{i=1}^{n} e_i v^i$. Show that A^{-1} is linear and $AA^{-1} = id_V$ and $A^{-1}A = id_{\mathbb{F}^n}$.

This means any vector space v of dimension $n < \infty$ is isomorphic to $\mathbb{F}^n$.

For instance, consider the vector space $P_2 = \{a_0 + a_1\theta + a_2\theta^2 | a_0, a_1, a_2 \in \mathbb{F}\}$ consisting of all polynomials of degree at most 2 with coefficients prove that $\{1, \theta, \theta^2\}$ is a basis for P_2 and thus P_2 has dimension 3. This means it is isomorphic to $\mathbb{F}^3$. Similarly, prove that P_n, the space of all polynomials in the variable θ with degree at most n has $\{1, \theta, \cdots, \theta^n\}$ as a basis and is thus isomorphic to $\mathbb{F}^n$.

Example : Consider $X = \sum_{i=1}^{\infty} X_i$ each $X_i = \mathbb{F}$. This is the space $\mathbb{F} \oplus \mathbb{F} \cdots$ (countably many copies). This can be regarded as the space of column vectors

$$x = \begin{pmatrix} x^1 \\ \cdot \\ \cdot \\ x^n \\ \cdot \\ \cdot \end{pmatrix}$$ of infinite depth; this amounts to saying that each $x \in X$ is an infinite sequence.

$\mathbb{N} \xrightarrow{x} \mathbb{F}$ with $x(i) = x^i \in \mathbb{F}$. It is much more convenient to think of this as the vector space of all polynomials in the variable θ, $X := \{a = a_0 + a_1\theta + \cdots + a_n\theta^n \mid 0 \leq n \in \mathbb{N}\}$; then show that $\{1, \theta, \theta^2, \cdots\}$ is a basis of X and thus this space is infinite-dimensional. (The polynomial $0 \in X$ is by agreement of degree $-\infty$; for this $a_i = 0$ for each i, recall that $a_0 + a_1\theta + \cdots + a_d\theta^d$ has degree d if $a_d \neq 0$). We denote this space by $\mathbb{F}[\theta]$.

Remark : In illustration above, it should be noted that a polynomial over $\mathbb{F}$ i.e. an element $a = a(\theta) \in \mathbb{F}[\theta]$ is not a polynomial function $\mathbb{F} \xrightarrow{\tilde{a}} \mathbb{F}$ given by $\tilde{a}(\lambda) := a_0 + a_1\lambda + \cdot s + a_n\lambda^n$. Indeed, suppose $\mathbb{F}$ is the field with only two elements 0 and 1 i.e. $\mathbb{F} = \mathbb{Z}_2$. The polynomial $a(\theta) = \theta + \theta^2 \in \mathbb{F}[\theta]$ is not the zero polynomial but (since $1+1 = 0$ in $\mathbb{Z}_2$) we have $\lambda + \lambda^2 = \tilde{a}(\lambda) = 0$ for each $\lambda \in \mathbb{F} = \mathbb{Z}_2$. Thus if $\tilde{\mathbb{F}}$ is the vector space of all polynomial functions $\mathbb{F} \to \mathbb{F}$ and we consider $\mathbb{F}[\theta] \xrightarrow{f} \mathbb{F}$ given by $f(a(\theta)) = \tilde{a}$, this is not an isomorphism in general. Prove however, that if $\mathbb{F}$ has an infinite number of elements (in particular if characteristic of $\mathbb{F}$ is 0) the function $f(a) := \tilde{a}$ defined by $a(\theta) = a_0 + a_1\theta + \cdots + a_n\theta^n \to \tilde{a}(\lambda) := a_0 + a_1\lambda + \cdots + a_n\lambda^n$ is an isomorphism.

Hint: A nonzero polynomial $a(\theta)$ of degree n has at most n zeros, so $\tilde{a}(\lambda) = 0$ is not possible for all λ. Thus no nonzero polynomial goes to the zero polynomial function. Prove that $f(a) := \tilde{a}$ is a linear transformation.

Example : Assuming that $\mathbb{F}$ has n + 1 distinct elements $\lambda_0, \cdots, \lambda_n$, consider the polynomials $L_i(\theta) := \Pi_{j \neq i} \dfrac{\theta - \lambda_j}{\lambda_i - \lambda_j}$ so that $L_i(\lambda_j) = \begin{cases} 1 & \text{if } i = j \\ 0 & \text{if } i \neq j \end{cases}$, $0 \leq i, j \leq n$. Given $a(\theta) = \sum_{i=0}^{n} \mu_i L_i(\theta)$ we have $a(\lambda_j) = \mu_j$ and thus a is the zero polynomial only when each $\mu_j = 0$. This means the set $\{L_0(\theta), \cdots, L_n(\theta)\}$ is linearly independent in the space P_{n+1} of polynomials in θ with degree at most n, and as such, spans a vector space of dimension n + 1 which must be then all of P_{n+1} since $\dim(P_{n+1}) = n + 1$. In other words, each polynomial $a(\theta) = a_o + \cdots + a_n\theta^n$ can be expressed uniquely as $a(\theta) = \sum_{i=0}^{n} a(\lambda_i) L_i(\theta)$; this formula is called the Lagrange interpolation formula.

(d) If S spans X then any linearly independent subset T or S can be completed to a basis $T \cup S_o$ for $<< S >>= X$ where $S_o \subseteq S$ and $S_o \cap T = \varphi$.

(e) In general, if P is a subspace of X, there are many subspaces Q such that $P \oplus Q = X$.

(f) If P and Q are finite dimensional subspaces of X, dim (P + Q) + dim (P ∩ Q) = dim P + dim Q. Where $P + Q := << P \cup Q >>$; in particular, dim (P ⊕ Q) = dim P + dim Q.

For instance, $\dim(\mathbb{F}^n) = \dim \mathbb{F} + \cdots + \dim \mathbb{F}$ (n times) $= 1 + \cdots + 1 = n$ (we have dim F = 1 since {1} clearly a basis because we can write any $\lambda \in \mathbb{F}$ as λ.1) because $\mathbb{F}^n = \mathbb{F} \oplus \cdots \oplus \mathbb{F}$ (n times).

Suppose $X \xrightarrow{A} Y$ is a linear transformation. Prove the following:

(i) The image of A (also called 'the image of X under A' and the 'range of A') $A(X) := \{A(x) \mid x \in X\}$ is a subspace of Y,

(ii) The kernel of A, ker $A := \{x \in X \mid A(x) = 0\}$, also denoted by $A^{-1}\{0\}$, is a subspace of X, and

(iii) A is injective iff ker A = {0}

Hint: (If ker A = {0} and Ax = Au, we have a(x – u) = A(x) – A(u) = 0 so that x – u = 0; if A is injective, we have A(x) = 0 = A(0) forcing x = 0).

(iv) If $\dim X = n < \infty$ with $\{e_1, \cdots, e_n\}$ as a basis and W is a vector space, an assignment $Ae_i = w_i \in W$ where $w_i \in W$ are arbitrarily chosen, will supply a unique linear transformation (also denoted by A) $X \longrightarrow W$ with the formula $A(x) := A(\sum_{i=1}^{n} x^i e_i) = \sum_{i=1}^{n} x^i A(c_i) = \sum_{i=1}^{n} x^i w_i$ for $x = \sum_{i=1}^{n} x^i e_i$.

(i) This sentence includes a claim with its proof. Extend this to an infinite-dimensional space

Hint: (The same proof works).

(ii) Use this to show that if dim X = n < ∞, and rank of A = ρ(A) := dim (A(X)), nullity of A = ν(A) := dim (ker A), we have ρ(A) + ν(A) = n = dim X.

Hint: Show that if ker A⊕Q = X then $Q \xrightarrow{B} \text{Im}\, A$ defined by Bq = Aq is an isomorphism; this is obvious since Im B = Im A and thus A is surjective and ker B = Q∩ker A = {0} i.e. B is injective. Now if $\{e_1, \cdots, e_q\}$ generate ker A, this set is linearly independent in X (prove this) and (∵<< X >>= X so that any linearly independent set in X can be extended to a basis of X by (d)) can be extended to a basis $\{e_1, \cdots, e_q, e_{q+1}, \cdots,$

$e_n\}$; clearly $<< e_{q+1}, \cdots, e_n >> \oplus \ker A = X$. As illustration, take $\mathbb{R}^2 \xrightarrow{A} \mathbb{R}^3$ defined by $A\begin{pmatrix} x^1 \\ x^2 \end{pmatrix} = \begin{pmatrix} x^1 \\ x^1 + x^2 \\ x^3 \end{pmatrix}$. Prove that it is a linear transformation and Im A has $\left\{\begin{pmatrix} 1 \\ 1 \\ 0 \end{pmatrix}, \begin{pmatrix} 0 \\ 1 \\ 1 \end{pmatrix}\right\}$ as a basis concluding that rank A = 2. Verify that nullity of A is 0. As another illustra tion, take $\mathbb{F}^3 \xrightarrow{A} \mathbb{F}^3$ given by $A\begin{pmatrix} x^1 \\ x^2 \\ x^3 \end{pmatrix} = \begin{pmatrix} x^1 - x^2 + 2x^3 \\ 2x^1 + x^2 \\ -x - 2x^2 + 2x^3 \end{pmatrix}$ observe that $\begin{pmatrix} a \\ b \\ c \end{pmatrix} \in \ker A$ iff $\frac{a}{-2} = \frac{b}{4} = \frac{c}{3}$ holds and prove that $\left\{\begin{pmatrix} -2 \\ 4 \\ 3 \end{pmatrix}\right\}$ is a basis for ker A so that A has nullity 1 and rank 2.

(a) As an exercise, show that $T^2 = 0$ ensures Im $T \subseteq \ker T$ but consider $\mathbb{F}^2 \xrightarrow{T} \mathbb{F}^2$ given by $T\begin{bmatrix} x^1 \\ x^2 \end{bmatrix} = \begin{bmatrix} x^2 \\ 0 \end{bmatrix}$ to conclude that $T^2 = 0$ need not mean $T = 0$.

(b) Moreover, prove that if T^2 and T have the same rank, for $X \xrightarrow{T} X$ then ker T = ker T^2 (they have the same dimension; we consider dim X < ∞) and ker T = {0}.

(c) Show that ker T = {0} iff each linearly independent subset of X is carried to a linearly independent subset by T. Also, prove that ker T = {0} iff $X \xrightarrow{T} W$ is an isomorphism (Now revisit the task at the end of the remark on page 6 where $\mathbb{F}[\theta] \xrightarrow{T} \tilde{\mathbb{F}}$ given by T (a(θ)) = ã) onto T (X).

(d) For dim X = dim Y < ∞, $X \xrightarrow{A} Y$ being linear, A is surjective ↔ A is an isomorphism.

Hint: (rank of A = dim Y iff nullity of A = 0 iff rank of A = dim X)

(e) Given $X_i \xrightarrow{B_i} Y_i \xrightarrow{A_i} Z_i$, i = i, 2 one can define $X_1 \oplus X_2 \xrightarrow{B_1 \oplus B_2} Y_1 \oplus Y_2 \xrightarrow{A_1 \oplus A_2} Z_1 \oplus Z_2$ with $(A_1 \oplus A_2).(B_1 \oplus B_2) = A_1B_1 \oplus A_2B_2$ and $(A_1 \oplus A_2) + (B_1 \oplus B_2) = (A_1 + B_1) \oplus (A_2 + B_2)$.

(Take $(A_1 \oplus A_2)(y_1 \oplus Y_2) = A_1(y_1) \oplus A_2(y_2)$ etc. Recall (f) saying P + Q :=<< P ∪ Q >>. Take everything finite dimensional for comfort).

Let X be a vector space and choose an indexed basis $e = \{e_i \mid i \in I\}$. If Y is another vector space and $A(e_i) \in Y$ are assigned i.e. if $e \xrightarrow{A} Y$ is a function, it extends uniquely to a linear transformation $X \xrightarrow{A} Y$ (again denoted by A) via the formula

$A(x) := \sum A(e_i)x^i$ for $\sum e_i x^i$. Choose $e \xrightarrow{e^i} \mathbb{F}$ by taking $e^i(e_j) = \begin{cases} 1 \text{ if } i-j \\ 0 \text{ if } i \neq j \end{cases}$ to produce linear form $X \xrightarrow{e^i} \mathbb{F}$ this acts on $x = \sum_k e_k x^k$ to produce $e^i(x) = x^i$. If $x \neq 0$ so that some $x^i \neq 0, e^i \in X'$ is non zero ($\because e^i(x) = x^i$) and thus for $X \xrightarrow{\Gamma} X''$ defined via the formula $< \Gamma(x) \mid \varphi > := < \varphi \mid x >$, we have $< \Gamma(x) \mid e^i > = < e^i \mid x > = x^i \neq 0$ i.e. $\Gamma(x)$ is nonzero. This means that $\ker \Gamma = \{o\}$. Since Γ was proved to be linear, we conclude that Γ is injective.

If $\dim X = n < \infty$, we find $< \varphi \mid x > = < \varphi \left| \sum_{i=1}^{n} e_i x^i \right. > = \sum_{i=1}^{n} < \varphi | e_i > x^i = \sum_{i=1}^{n} < \varphi | e_i > < e_i | x >$.

This holds at each $x \in X$ so we have $\varphi = \sum_{i=1}^{n} \varphi_i e^i$ where $\varphi_i = < \varphi | e_i > \in \mathbb{F}$. In other words, each $\varphi \in X'$ can be written uniquely as a linear combination of the forms $e^i \in X', 1 \leq i \leq n$. This means $\{e^i \mid 1 \leq i \leq n\}$ is a basis for X' and we conclude $X = << e_1, \cdots, e_n >> \Rightarrow X' = << e^1, \cdots, e^n >>$.®

But this means $\dim X = \dim X' = n$ and then $\dim X'' = n$. Thus the injective linear $X \xrightarrow{\Gamma} X''$ must be an isomorphism. We identify therefore $x \in X$ with $\Gamma(x) \in X''$ which is given by $< \Gamma x \mid \varphi > = < \varphi \mid x >$, in particular, $\Gamma(e_i)$ with e_i. This permits us to write $< \varphi \mid x >$ also as $< x \mid \varphi >$ (i.e. $\varphi(x) \in \mathbb{F}$ as $x(\varphi)$) for $x \in X, \varphi \in X'$ and we do it beginning now. The implication ® now exhibits an isomorphism $X \cong X'$ via $e_i \to e^i$ which is involutive in the same sense that $X'' = X'$ and (prove this) $A'' = A$ for $X \xrightarrow{A} Y$.

Exercise

Prove that for $X \xrightarrow{A} Y \xrightarrow{B} Z$ we have $(BA)' = A'B'$ and that $(id_X)' = id_X$.

Hint:

Just use $< A'(\psi) \mid x > : - < \psi \mid A(x) >$, the defining formula for A'.

Example 0.4 Essentially we know that there is only one finite-dimensional vector space of dimension n; this is $\mathbb{F}^n$ with basis $e_i = \begin{pmatrix} 0 \\ 1 \\ 0 \end{pmatrix} \to$ i-th place. Then while $(\mathbb{F}^n)'$ is (isomorphic to) $\mathbb{F}^n$, we write its elements as row vectors $x = [x_1, \cdots, x_n]$. At present, it may be convenient to take $\varphi \in (\mathbb{F}^n)'$ as $\varphi = [\varphi_1, \cdots, \varphi_n]$, $\varphi_i \in \mathbb{F}$, $\varphi = \sum_{i=1}^{n} \varphi_i e^i$ (where of course $e^i = [0, \cdots, 1, 0, \cdots]$)

for notational consistency. Then, for $x = \begin{pmatrix} x^1 \\ \cdot \\ \cdot \\ \cdot \\ x^n \end{pmatrix} \in X \cong \mathbb{F}^n$ and $\varphi = [\varphi_1, \cdots, \varphi_n] \in X' \cong \mathbb{F}^n$,

we have $< \varphi \mid x >= [\varphi_1, \cdots, \varphi_n] \begin{pmatrix} x^1 \\ \cdot \\ \cdot \\ \cdot \\ x^n \end{pmatrix} = \sum_{i=1}^{n} \varphi_i x^i \in \mathbb{F}$ (but don't say 'dot product'; that is still in distant future).

That is, the action of the row vector φ which is a linear form on $\mathbb{F}^n$ on the column vector x which is a vector in $\mathbb{F}^n$ is the matrix product of φ with x.

Exercise

For $X \xrightarrow{A} Y$, prove that

(a) ker $A' = (\text{Im } A)^o$, (b) Im $A' = (\ker A)^o$ and rank $A' =$ rank A', nullity A = nullity A'. (Ofcourse, for dim $X < \infty$, dim $Y < \infty$), where we define the annihilator P^o of a subset P $\subseteq X$ by $P^o := \{\varphi \in X' \mid < \varphi \mid p >= 0 \text{ at each } p \in P\}$.

(Then prove that P^o is a subspace of X' regardless of whether P is or is not a subspace of X (Just note that $< \lambda\varphi + \varsigma \mid p >= \lambda < \varphi \mid p > + < \varsigma \mid p >= 0$ for $\varphi, \varsigma \in X'$).

Now if $<< X >>=<< e_1, \cdots, e_n >>$ and $W =<< e_1, \cdots, e_r >>$ is a subspace of X we clearly have $W^o =<< e^{r+1}, \cdots, e^n >>$$(< \varphi \mid p >=< \sum \varphi_i e^i \left| \sum_{i=1}^{r} e_i p^i \right> = \sum_{j=1}^{n} \sum_{i=1}^{r} \varphi_j < e^j | e_i > p^i = 0$ demands that $\varphi \in << e^{r+1}, \cdots, e^n >>$). Since $<< e^{r+1},\ \ , e^n >>$, we conclude that dim W + dim W^o = dim X. It should be also clear that $W^{oo} = W$.

Then $< A'(\psi) \mid x >=< \psi Ax >$ says that $\psi \in \ker A'$ iff $< \varphi \mid Ax >= 0$ at each $x \in X$ i.e. iff $\varphi \in (\text{Im } A)^o$. Recalling that $W^{oo} = W$, if we use A' for A, we have ker A = ker $A'' = (\text{Im } A')^o$ which means $(\ker A)^o = (\text{Im } A')^{oo} = \text{Im } A'$. Since nullity of A' is dim (ker A') = dim $(\text{Im } A)^o$ and since dim Y' = rank (A') + nullity A' = dim (Im A') + dim $(\text{Im } A)^o$ = dim (Im A) + dim $(\text{Im } A)^o$ we have rank A = rank A', similarly, nullity A = nullity A').

Matrices and Linear Transformations

We work with a linear transformation $X \xrightarrow{A} Y$ and remind ourselves that for φ(x) ∈ $\mathbb{F}$ is < φ | x >. Then | Y >< φ stands for the function X → Y which acts on x ∈ X by returning | y > < φ | x > where we write

(i) | x >, | y > etc. for vectors x, y in X and Y etc. calling them 'ket's and < ϕ |, < ψ |, etc. for forms ϕ, ψ in X' and Y' etc. calling them 'bra's so that < ϕ | x > is a bar(c)ket ∈ $\mathbb{F}$ and | x > < ϕ | is a ket-bra which is a linear transformation X → X returning | x >< ϕ | w >∈ X on input | w >∈ X.

This is Dirac notion and terminology for linear algebra.

If $X \xrightarrow{|y><\varphi|} Y$ is not zero then

(i) $\text{Im}(\|y><\varphi\|) =<< y >>$ ($\because$ $v \in \text{Im}(\|y><\varphi\|)$ iff $v = \|y><\varphi\|x>$ for some x ∈ X i.e. a scalar multiple of $\|y>$; note that since we have here $<\varphi\| \neq 0$ we have $0 \neq \varphi_i =<\varphi\|e_i>$

for some basis vector e_i and then any λ ∈ $\mathbb{F}$ is < φ | e_i > for the choice $x = \frac{e_i\lambda}{\varphi_i}$ since then

$$<\varphi\Big|x>=<\varphi\Big|\frac{e_i\lambda}{\varphi_i}>=\frac{<\varphi|e_i>}{\varphi_i}\lambda=\lambda \text{ so that } y\lambda=\Big|y><\varphi\Big|\frac{e_i\lambda}{\varphi_i}>$$ for each yλ in << y >>).

(ii) ker (| y > < φ |) =<< φ >>° (recall X" = X) (x ∈ ker (| y > < φ |) iff | y > < φ | x >= o iff < φ | x >= o since | y >≠ o. This means, since o =< φ | x >=< x | φ >, that x ∈<< φ >>°).

(iii) Bilinearity: | λy + v > < φ | x >=| λy > < φ | x > + | v > < φ | x >= [λ | y > < φ | + | v > < φ](x) at each x ∈ X so that | λy + v > < φ |= λ | y > < φ | + | v > < φ |, and |> < φ + ς μ | x >=| y > < φ | x > + | y > < ς | x > μ = [| y > < φ | +μ | y > < ς |](x) at each x, so | y > < φ + ς μ |=| y > < φ | + |> < ς | μ for each y, v ∈ Y , φ, ς ∈ X' , λ, μ ∈ $\mathbb{F}$.

(iv) For $W \xrightarrow{A} X \xrightarrow{|><\varphi|} Y \xrightarrow{B} Z$ we have $B \circ (|y><\varphi|x>=|By><\varphi|x>$ at each

$x \in X$ so that $B \circ |y><\varphi| = |By><\varphi|(X \to X)$ and (∵ < φ | Aw >= ((| y >< φ |) ∘ A) w at each w ∈ X so that (| y >< φ |) ∘ A =| y >< $A'\varphi'$ | (W → Y).

(v) Since < A' ψ |x > = < ψ| Ax > for $X \xrightarrow{A} Y$, we have < (|y > < φ|) ' ψ| x > = < ψ| |y > < φ |x > at each x ∈ X which means (|y > < φ|) ' (ψ) =< ψ| |y > < φ|> =|ψy > < φ| = (|φ > < y|)(ψ) =|φ > < y| ψ > (we have $X \xrightarrow{|y><\varphi|} X \xrightarrow{\psi} \mathbb{F}$, then $\psi \circ |y><\varphi| = |\psi(y)><\varphi|$ but $\psi(y) =<\psi|y>=<y|\psi>$ since y ∈ Y = Y" and thus ψ(y)(φ) ∈ X' is simply (< y | ψ >)φ which we are writing as | φ >< y | ψ >; note that < y | ψ >=< ψ | y >∈ $\mathbb{F}$) at each ψı Y'.

Thus we conclude:

$(|>\,<\varphi|)' = |\varphi><y|$ where $|y>\in Y$ on LHS has been written as $<y|\in Y''$ while $<\varphi|\in X'$ on LHS has been written as $|\varphi>\in X'$ as a ket since we have $Y' \xrightarrow{(|y><\varphi|)'=|><y|} X'$ just as we have $Z \xrightarrow{|w><\alpha|} W$ for $|w>\in Y$, $<\alpha|\in Z'$.

(vi) For $X \xrightarrow{|y><\varphi|} Y \xrightarrow{|z><\psi|} Z$

$(|Z><\psi|)\circ(|y><\varphi|) = <\psi|y>|z><\varphi|$ where $y\in Y$, $\psi\in Yt$, $\varphi\in X'$ and $x\in Z$ (use $B\circ(|y><\varphi|) = |By><\varphi|$ established in (iv) with $B=|z><\psi|$, noting that $<\psi|y\in\mathbb{F}$).

(vii) For $X = << e_1,\cdots,e_n >>$ so that $X' = << e',\cdots,e' >>$, we have $\sum_{i=1}^{n}|e_i><e^i|e_j> = |e_j>$

which means $id_X = \sum_{i=1}^{n}|e_i><e^i|$ (because $\sum_{i=1}^{n}|e_i><e^i|$ is seem $e\to X$ given by $e_j\to e_j$) and $\sum_{i=1}^{n}|e_i><e_i|e^j> = |e^j>$ ($\because <e_i|e^j> = <e^j|e_i> = \delta_{ij} == \begin{cases}1 \text{ if } i=j\\ 0 \text{ if } i\neq j\end{cases}$) so that $id'_X = \sum_{i=1}^{n}|e^i><e_i|$.

(viii) Suppose $X = << e_1,\cdots,e_n >>$, $Y = << d_1,\cdots,d_m >>$. Given $X \xrightarrow{A} Y$ we have $A(e_i)\in Y$ and can write $A(e_i) = \sum_{j=1}^{m}|d_j>a_i^j$ for scalar $a_i^j\in\mathbb{F}$ uniquely; $a_i^j = <d^j|Ae_i>$.

Then $A = A\circ id_X = A\circ(\sum_{i=1}^{n}|e_i><e^i|$ (see (vii) above) $= \sum_{i=1}^{n}|Ae_i><e^i| = \sum_{i=1}^{n}\sum_{j=1}^{m}|d_j>a_i^j<e^i| = \sum_{i=1}^{n}\sum_{j=1}^{m}a_i^j|d_j><e^i|$

which means that the m.n linear transformation $X \xrightarrow{|d_j><e^i|} Y$ which act on $x\in X$ to return the vector $|d_j><e^i|x> = |d_j>X^i$ form a basis for the space L(X, Y) since we just saw that any $X \xrightarrow{A} Y$ can be written uniquely as a linear combination $A = \sum_{i=1}^{n}\sum_{j=1}^{m}a_i^j|d_j><e^i|$ in terms of these $|d_j><e^i|$. The m.n scalars a_i^j are written as an $m\times n$ matrix $a = [a_i^j]$ with i indexing the columns and j indexing the rows and this is called the matrix associated with the linear transformation $X \xrightarrow{A} Y$ with reference to the given indexed bases $e = \{e_1,\cdots,e_n\}$ of X and $d = \{d_1,\cdots,d_m\}$ of Y.

Just how do we write this matrix? To begin, recall that when we say $y\in Y$ is uniquely written as $\sum_{j=1}^{m}d_jy^j$ the scalars y^j are $<d^j|y>\in\mathbb{F}$ where $\{d_1,\cdots,d_m\}$ is the basis of Y' associated with the basis $d = \{d_1,\cdots,d_m\}$ defined by $<d^j|d_k> := \delta_{jk} = \begin{cases}1 \text{ if } j=k\\ 0 \text{ if } j\neq k\end{cases}$.

Here $A(e_i) \in Y$ and so there scalars $(A(e_i))^j$ are $< dj \mid Ae_i >$ which we have written as a_i^j; they depend on both $\{e_1, \cdots, e_n\}$ and $\{d_1, \cdots, d_m\}$ (and of course on $\{e^1, \cdots, e^n\}$ and $\{d^1, \cdots, d^n\}$).

The matrix $a = \left[a_i^j\right]$ is obtained by writing the m-deep columns $A(e_i)$ as the i-the column and

thus $a = \begin{pmatrix} a_1^1 & a_2^1 & \cdots & \cdots & a_n^1 \\ a_1^2 & a_2^2 & \cdots & \cdots & a_n^2 \\ \cdot & \cdot & \cdot & \cdot & \cdot \\ \cdot & \cdot & \cdot & \cdot & \cdot \\ \cdot & \cdot & \cdot & \cdot & \cdot \\ a_1^m & a_2^m & \cdots & \cdots & a_n^m \end{pmatrix}$ with a_i^j at the intersection of the j-th row and the i-the column.

(ix) The discussion in (viii) will be summarized by writing $\left(X, \underline{\underline{e}}\right) \xrightarrow{a=\left[a_i^j\right]} \left(Y, \underline{\underline{\alpha}}\right)$ and we keep writing $a_i^j =< d^j \middle| Ae_i >$ so that we never forget that matrix representation is in terms of the indexed bases $\underline{\underline{e}}$ and $\underline{\underline{d}}$ (and ofcourse their 'dual' or 'reciprocal' bases $\{e^1, \cdots, e^n\}$, $\{d^1, \cdots, d^m\}$) this is important because if the indexing is changed, the location of a_i^j is charged in the matrix and if either of the bases is changed, the entries will be different.

If $\left(X, \underline{\underline{e}}\right) \xrightarrow[A]{a=\left[a_i^j\right]} \left(Y, \underline{\underline{d}}\right) \xrightarrow[B]{b=\left[b_j^k\right]} \left(Z, \underline{\underline{c}}\right)$, $1 \le i \le n$, $1 \le j \le m$, $1 \le k \le p$, we have

$(b \circ a)_1^k =< c^k \left|(B \circ A)e_i >=< C^k \right| B(\sum_{j=1}^{m} a_i^j \left| d_j >) >= \sum_{j=1}^{m} < C^k B d_j > a_i^j = \sum_{j=1}^{m} b_j^k a_i^j$ so that the composition of the linear transformation $B \circ A$ corresponds exactly to the composition of their matrix representation in the given bases.

Further, since $Y' =<< d^1, \cdots, d^m >>$ and $X' =<< e^1, \cdots, e^n >>$, we have the matrix representation of A' given by $Y' \xrightarrow[a']{A'} X'$ where $(a')_j^i =< e_i \left| A'(d^j) \right> =< A(e_i) \middle| d^j > =< d^j \middle| A(e_i) >= a_i^j$.

(Note that we use $Y'' = Y$, $X'' = X$, and the fact that $< \varphi \mid x >=< x \mid \varphi >$ for kets $\mid x >$ and bras $< \varphi \mid$).

Also, $\left(X, \underline{\underline{e}}\right) \xrightarrow[I]{id_X} \left(X, \underline{\underline{e}}\right)$ has $I_i^e =< e^l \middle| id_X(e_i) >=< e^l \middle| e_i > = \begin{cases} 1 \text{ if } l = i \\ 0 \text{ if } l \ne i \end{cases}$, $1 \le i \le n$, $1 \le l \le n$; thus the identity operator with respect to the same basis has the representation given by the identity(square) matrix which has on-diagonal entries 1 and off-diagonal entries 0.

We shall now supply some examples to illustrate the discussion. We use the standard terminology: Choice of a basis $\underline{\underline{e}}$ for X is called a coordinatization of X, in the representation $x = \sum e_i x^i = \sum \left| e_i > x^i\right.$, the scalar x^i are called the coordinates or the components of the vector x with respect to the given coordinatization, and a linear transformation $X \xrightarrow{A} X$ will be frequently called a 'linearoperator on X'.

Some Illustrations

Many authors do not wish to emphasize the need for indexing a basis, at least in the beginning. There is a reason for it. Consider the basis $\left\{e_1 = \begin{bmatrix}1\\0\end{bmatrix}, e_2 = \begin{bmatrix}0\\1\end{bmatrix}\right\}$ for $\mathbb{F}^2$; the indexing would mean that $\{e_2, e_1\}$ is not a basis ($\{e_1, e_2\}$ but a different basis (for this reason, ordered basis is usually the term for what we are calling indexed basis). To describe the action of a linear operator $\mathbb{F}^2 \xrightarrow{A} \mathbb{F}^2$ on basis vectors, one could say either

(i) $Ae_1 = \alpha e_1$, $Ae_2 = \beta e_1 + \gamma e_2$ (in which ordering the basis is irrelevant), or (ii) A has matrix $\begin{bmatrix}\alpha & \beta\\ 0 & \gamma\end{bmatrix}$ with respect to the ordered basis $\{e_1, e_2\}$ (in which case ordering the basis is crucial, since one could easily but of course mistakenly, take this to mean $Ae_1 = \alpha e_1 + \beta e_2$, $Ae_2 = \gamma e_2$). The best thing to do seems to be: keep writing "$a_i^j = < d^j | Ae_i >, 1 \le j \le m\ 1 \le i \le n$ and $a = \left[a_i^j\right]$ is an $m \times n$ matrix for $(X, \underline{\underline{e}}) \xrightarrow{A} (Y, \underline{\underline{d}})$: $Ae_i = \sum |d_j > a_i^j$." Linear algebra is better understood without matrices but the subject is so computational in application that de-emphasizing matrices is almost scandalous.

(i) Find the matrix of $\mathbb{F}^3 \xrightarrow{A} \mathbb{F}^2$ given by $A\begin{pmatrix}x\\y\\z\end{pmatrix} = \begin{pmatrix}2x+3y-z\\4x-y+2z\end{pmatrix}$ relative to the basis

$$\left\{\underbrace{\begin{pmatrix}1\\1\\0\end{pmatrix}}_{=b_1}, \underbrace{\begin{pmatrix}1\\2\\3\end{pmatrix}}_{=b_2}, \underbrace{\begin{pmatrix}1\\3\\5\end{pmatrix}}_{=b_3}\right\} \text{ of } \mathbb{F}^3 \text{ and } \left\{\underbrace{\begin{pmatrix}1\\2\end{pmatrix}}_{=c_1}, \underbrace{\begin{pmatrix}2\\3\end{pmatrix}}_{=c_2}\right\} \text{ of } \mathbb{F}^2.$$

Hint:

That the two sets given are indeed bases can be verified by checking them for linear independence. For instance, we know that $\mathbb{F}^3$ has three elements in any basis and the system $\lambda_1 b_1 + \lambda_2 b_2 + \lambda_3 = o$ has no nontrivial solutions (verify) so that $\{b_1, b_2, b_3\}$ are linearly independent and must form a basis of $\mathbb{F}^3$. Next, if $\begin{pmatrix}a\\b\end{pmatrix} \in \mathbb{R}^2$, solving $x\begin{pmatrix}1\\2\end{pmatrix} + y\begin{pmatrix}2\\3\end{pmatrix} = \begin{pmatrix}a\\b\end{pmatrix}$ supplies $x = -3a + 2b$, $y = 2a - b$ so that in the given basis $\{c_1, c_2\}$ of $\mathbb{F}^2$, we must have this vector as $\begin{pmatrix}-3a+2b\\2a=b\end{pmatrix}$ and directly calculating therefore, we have

$Ab_1 = \begin{pmatrix}5\\3\end{pmatrix} = -9c_1 + 7c_2, Ab_2 = c_1 + 2c_2, Ab_3 = 4c_1 + c_2$. The desired matrix will be obtained by these column vectors Ab_1, Ab_2, Ab_3 so that it is $a = \begin{pmatrix} -9 & 1 & 4 \\ 7 & 2 & 1 \end{pmatrix}$. The vector $\begin{pmatrix} 2 \\ 5 \\ -3 \end{pmatrix}$ of $\mathbb{F}^3$ is calculated to be $\begin{pmatrix} 11 \\ -21 \\ 12 \end{pmatrix}$ in the basis $\{b_1, b_2, b_3\}$ (verify that any $\begin{pmatrix} a \\ b \\ c \end{pmatrix} \in \mathbb{F}^3$ in the basis $\{b_1, b_2, b_3\}$ will be $\begin{pmatrix} -a+2b-c \\ 5a-5b+2c \\ -3a+3b-3c \end{pmatrix}$) and calculating $A\begin{pmatrix} 2 \\ 5 \\ -3 \end{pmatrix} = \begin{pmatrix} 4+15+3 \\ 8-5-6 \end{pmatrix} = \begin{pmatrix} 22 \\ -3 \end{pmatrix}$ by the supplied formula, writing this as $\begin{pmatrix} -66-6 \\ 44+3 \end{pmatrix}$ is the basis $\{c_1, c_2\}$, we get $\begin{pmatrix} -72 \\ 47 \end{pmatrix}$ which is precisely $\begin{pmatrix} -9 & 1 & 4 \\ 7 & 2 & 1 \end{pmatrix}\begin{pmatrix} 11 \\ -21 \\ 12 \end{pmatrix}$ as it should be and displays the action of A as matrix action.

(ii) For $\mathbb{F}^3 \xrightarrow{A} \mathbb{F}^2$ given by $A\begin{pmatrix} x \\ y \\ z \end{pmatrix} = \begin{pmatrix} 3x+2y-4z \\ x-5y+3z \end{pmatrix}$ with $B_1 = \left\{ \underbrace{\begin{pmatrix} 1 \\ 1 \\ 1 \end{pmatrix}}_{=u_1}, \underbrace{\begin{pmatrix} 1 \\ 1 \\ 1 \end{pmatrix}}_{=u_2}, \underbrace{\begin{pmatrix} 1 \\ 1 \\ 1 \end{pmatrix}}_{=u_3} \right\}$ and $B_2 = \left\{ \underbrace{\begin{pmatrix} 1 \\ 3 \end{pmatrix}}_{=v_1}, \underbrace{\begin{pmatrix} 2 \\ 5 \end{pmatrix}}_{=v_2} \right\}$ as bases, the representation of A is $\begin{pmatrix} -7 & -33 & -13 \\ 4 & 19 & 8 \end{pmatrix}$. Verify this and the action displayed on $\begin{pmatrix} x \\ y \\ z \end{pmatrix}$ to be $\begin{pmatrix} -13x-20y+26z \\ 8x+11y-15z \end{pmatrix}$ in the coordinatizon B_2 for $A\begin{pmatrix} x \\ y \\ z \end{pmatrix}$ as given.

Prove that if $X \xrightarrow{A} Y$ is a linear transformation, there exists a basis $\underline{\underline{e}}$ for X and a basis $\underline{\underline{d}}$ for Y for which A has the matrix representation. $a = \begin{pmatrix} I_r & 0 \\ 0 & 0 \end{pmatrix}$ where I_r is the identity matrix of order r and r is the rank of A; take dim X = n < ∞, dim Y = m < ∞.

Hint:

We know that $r = \dim(A(\alpha))$ hence dim (ker A) = n – r). Let $\{e_{r+1}, \cdots, e_n\}$ be a basis for ker A then as a linearly independent subset of X, this can be extended to a basis $\underline{\underline{e}} = \{e_1, \cdots e_r, e_{r+1}, \cdots, e_n\}$ of X. Write $d_j = A(e_j)$ for $1 \le j \le r$. Since $X \xrightarrow{A} A(X)$ is surjective, it must be an isomorphism $X \xrightarrow{A} A(X)$ (d) and thus these r vectors $d_j \in A(X)$ will form a basis for the r-dimensional space A(X); then they are linearly independent in Y and we can extend it to a basis $\underline{\underline{d}} = \{d_1, \cdots, d_m\}$ of Y (read the HINT for (ii) if you are uncertain about the process of getting $\underline{\underline{e}}$ and $\underline{\underline{d}}$). Then

$d_1 = A(e_1) = o.d_1 + o.d_2 + \cdots + o.d_m$

$d_2 = A(e_2) = o.d_1 + 1.d_2 + \cdots + o.d_m$

… … …

$d_r = A(e_r) = o.d_1 + \cdots + 1.d_r + o.d_{r+1} \cdots + o.d_m$

$o = A(e_{r+1}) = o.d_1 + \cdots + o.d_m$ ($\because e_{r+1} \in \ker A$)

… … …

$o = a(e_n) = o.d_1 + \cdots + o.d_m$ ($\because e_n \in \ker A$)

and therefore $(X, \underline{\underline{e}}) \xrightarrow[a]{A} (Y, \underline{\underline{d}})$ must be $a = \begin{pmatrix} I_r & 0 \\ 0 & 0 \end{pmatrix}$ as advertized.

What is dim (L(X, Y)) if dim X = n < ∞, dim Y = m < ∞?

Hint:

Fix a basis $\underline{\underline{e}} = \{e_1, \cdots, e_n\}$ for X and a basis $\underline{\underline{d}} = \{d_1, \cdots, d_m\}$ for Y. Then the corresponding $(X, \underline{\underline{e}}) \xrightarrow[a]{A} (Y, \underline{\underline{d}})$ $A \leftrightarrow a$, is bijective. Since the vector space of all m × n matrices (over $\mathbb{F}$) is of dimension m.n (with basis $\begin{pmatrix} 1 & 0 & \cdots & 0 \\ 0 & 0 & \cdots & 0 \\ 0 & 0 & \cdots & 0 \end{pmatrix}, \begin{pmatrix} 0 & 1 & 0 & \cdots \\ 0 & \cdots & \cdots & \cdots \end{pmatrix}, \begin{pmatrix} 0 & \cdots & 0 \\ 1 & \cdots & \cdots \\ 0 & \cdots & \cdots \end{pmatrix}$

etc; prove this), dim (L(X, Y)) = m.n.

But it is far better to simply appeal to the fact that $\{|d_j\rangle\langle e_i| \; | 1 \le i \le n \; 1 \le j \le m\}$ forms a basis for L(X, Y).

If $X \times X \xrightarrow{B} \mathbb{F}$ is such that B(λx + u, w) = λB(x.w) + B(u, w) and B(x, u + wλ) = B(x, u) + B(x, w)λ, we say it is a bilinear form on X. Show that if B and C are bilinear forms on X, (B + C) (x, u) := B(x, u) + C(x, u) makes the collection of all bilinear forms on X a vector space(with the obvious scalar multiplication). Call this $\underline{\underline{\mathrm{Bi\,lim}(X)}}$.

(i) If $\underline{\underline{e}}' = \{e_1', \cdots, e_n'\}$ is a basis of X' show that $\beta_i^j(x, u) := \beta_j'(x)\beta_i'(u)$ forms a basis of $\underline{\underline{\mathrm{Bi\,lim}(X)}}$.

Hint:

If $B \in \underline{\underline{\mathrm{Bi\,lim}(X)}}$ and $b_i^j := B(e_j, e_i) \in \mathbb{F}$ where $\underline{\underline{e}} = \{e_1, \cdots, e_n\}$ is the corresponding basis for $X = X''$, we get

$(\sum_{i,j} b_i^j \beta_i^j)(e_s, e_t) = \sum_{i,j} b_i^j \beta_i^j)(e_s, e_t) = \sum_{i,j} b_i^j e_i^{'}(e_s) e_i^{'}(e_t) = \sum b_i^j \delta_{js} \delta_{it} = b_t^s = B(e_s, e_t)$; since $\{(e_s, e_t) \mid 1 \le s \le n, 1 \le t \le n\}$ forms a basis of X × X, we get all of B from this; thus the β_i^j span $\underline{\underline{\text{Bi lim}(X)}}$ (verify of course that β_i^j are bilinear). If $B = \sum b_i^j \beta_i^j = 0$, we have $B(e_s, e_t) = b_s^t = 0$ for each s, t which means this set is linearly independent and is thus a basis for $\underline{\underline{\text{Bi lim}(X)}}$. In particular dim ($\underline{\underline{\text{Bi lim}(X)}}$= (dim X)².

(ii) Entering $b_i^j = B(e_j, e_j)$ into n² matrix at the intersection of the j-th row and i-th column will get us a matrix b which is called the matrix of the bilinear form B with respect to the basis $\underline{\underline{e}}$ of X.

For example, the bilinear form on $\mathbb{F}^2$ given by $B\left(\begin{pmatrix} x_1 \\ x_2 \end{pmatrix}, \begin{pmatrix} y_1 \\ y_2 \end{pmatrix}\right) := 2x_1y_1 - 3x_1y_2 + x_2y_2$,

with respect to the basis $\left\{ e_1 = \begin{pmatrix} 1 \\ 0 \end{pmatrix}, e_2 = \begin{pmatrix} 1 \\ 1 \end{pmatrix} \right\}$ has the matrix $\begin{pmatrix} 2 & -1 \\ 2 & 0 \end{pmatrix}$ (Verify this).

(iii) Given a Bilinear form $X \times X \xrightarrow{B} \mathbb{F}$ on X, the function $X \xrightarrow{Q} \mathbb{F}$ supplied by Q(x):= B (x, x) is called the quadratic form associated to B.

Show that the correspondence established by (ii) above, B ↔ b is bijective in the sense that B(x, u) = x^tbu (x^t is the transpose of the column vector x) and that any n²-matrix b will raise a bilinear form B(x, u):= x^tbu. Further, show that Q ↔ q establishes a similar correspondence between quadratic forms Q and symmetric n²-matrices q via Q(x, x):= x^tqx.

(iv) Show that if B raises the quadratic form Q, we can get the bilinear form B from

2B (u, v) = B (u, u) + B (u, v) + B (v, u) + B (v, v) – B (u, u) – B (v, v)

= B (u + v, u + v) – B (u, u) – f (v, v)

= Q (u + v) – Q (u) – Q (v)

that is, B(u, v) can be defined from the quadratic form Q via

$B(u,v) := \frac{1}{2}\left[Q(u+v) - Q(u) - Q(v)\right]$ (provided of course that $2 \ne 0$ in $\mathbb{F}$ i.e. $\mathbb{F} \ne \mathbb{Z}_2$).

Note:

Frequently, to save space $\begin{pmatrix} x^1 \\ x^2 \end{pmatrix} \in \mathbb{F}^n$ will be written $[x^1, \cdots, x^n]^t$ or $(x^1, \cdots, x^n)$.

Cauchy–Schwarz Inequality

In mathematics, the Cauchy–Schwarz inequality, also known as the Cauchy–Bunyakovsky–Schwarz inequality, is a useful inequality encountered in many different settings, such as linear algebra, analysis, probability theory, vector algebra and other areas. It is considered to be one of the most important inequalities in all of mathematics. It has a number of generalizations, among them Hölder's inequality.

The inequality for sums was published by Augustin-Louis Cauchy (1821), while the corresponding inequality for integrals was first proved by Viktor Bunyakovsky (1859). The modern proof of the integral inequality was given by Hermann Amandus Schwarz (1888).

Statement of the Inequality

The Cauchy–Schwarz inequality states that for all vectors u and v of an inner product space it is true that

$$|\langle u, v\rangle|^2 \leq \langle u, u\rangle \cdot \langle v, v\rangle$$

where $\langle \cdot, \cdot \rangle$ is the inner product. Examples of inner products include the real and complex dot product. Equivalently, by taking the square root of both sides, and referring to the norms of the vectors, the inequality is written as

$$|\langle u, v\rangle| \leq \| u \| \ \| v \|.$$

Moreover, the two sides are equal if and only if **u** and **v** are linearly dependent (meaning they are parallel, one of the vector's magnitudes is zero, or one is a scalar multiple of the other).

If $u_1, \ldots, u_n \in \mathbb{C}$ and $v_1, \ldots, v_n \in \mathbb{C}$ have an imaginary component, the inner product is the standard complex inner product where the bar notation is used for complex conjugation and then the inequality may be restated more explicitly as

$$|u_1\bar{v}_1 + \cdots + u_n\bar{v}_n|^2 \leq (|u_1|^2 + \cdots + |u_n|^2)(|v_1|^2 + \cdots + |v_n|^2)$$

or

$$\left|\sum_{i=1}^{n} u_i \bar{v}_i\right|^2 \leq \sum_{j=1}^{n} |u_j|^2 \sum_{k=1}^{n} |v_k|^2.$$

Proofs

First Proof

Let u and v be arbitrary vectors in a vector space over $\mathbb{F}$ with an inner product, where $\mathbb{F}$ is the field of real or complex numbers. We prove the inequality

$$|\langle u, v\rangle| \leq \| u \| \ \| v \|$$

and that equality holds if and only if either u or v is a multiple of the other (which includes the special case that either is the zero vector).

If $v = 0$, it is clear that we have equality, and in this case u and v are also linearly dependent, regardless of u, so the theorem is true. Similarly if $u = 0$. We henceforth assume that V is nonzero.

Let

$$z = u - u_v = u - \frac{\langle u, v\rangle}{\langle v, v\rangle} v.$$

Then, by linearity of the inner product in its first argument, one has

$$\langle z, v\rangle = \left\langle u - \frac{\langle u, v\rangle}{\langle v, v\rangle} v, v \right\rangle = \langle u, v\rangle - \frac{\langle u, v\rangle}{\langle v, v\rangle} \langle v, v\rangle = 0.$$

Therefore, z is a vector orthogonal to the vector v (Indeed, z is the projection of u onto the plane orthogonal to v.) We can thus apply the Pythagorean theorem to

$$u = \frac{\langle u, v\rangle}{\langle v, v\rangle} v + z$$

which gives

$$\| u \|^2 = \left| \frac{\langle u, v\rangle}{\langle v, v\rangle} \right|^2 \| v \|^2 + \| z \|^2 = \frac{|\langle u, v\rangle|^2}{\| v \|^2} + \| z \|^2 \geq \frac{|\langle u, v\rangle|^2}{\| v \|^2}$$

and, after multiplication by $\| v \|^2$ and taking square root, we get the Cauchy–Schwarz inequality. Moreover, if the relation $\geq$ in the above expression is actually an equality, then $\| z \|^2 = 0$ and hence $z = 0$; the definition of z then establishes a relation of linear dependence between u and v. On the other hand, if u and v are linearly dependent, then there exists $c \in \mathbb{F}$ such that $u = c \cdot v$ (since $v \neq 0$). Then

$$|\langle u, v\rangle| = |\langle c \cdot v, v\rangle| = \left| c \| v \|^2 \right| = |c| \| v \|^2 = \| c \cdot v \| \ \| v \| = \| u \| \ \| v \|.$$

This establishes the theorem.

Second Proof

Let u and v be arbitrary vectors in a vector space $\mathbb{F}$ with an inner product, where $\mathbb{F}$ is the field of real or complex numbers.

In the special case $v = 0$ the theorem is trivially true. Now assume that $v \neq 0$. Let $\lambda \in \mathbb{C}$ be given by $\lambda = \langle u, v\rangle / \| v \|^2$, then

$$
\begin{aligned}
0 &\le \| u - \lambda \cdot v \|^2 \\
&= \langle u,u\rangle - \langle \lambda \cdot v, u\rangle - \langle u, \lambda \cdot v\rangle + \langle \lambda \cdot v, \lambda \cdot v\rangle \\
&= \langle u,u\rangle - \lambda\langle v,u\rangle - \bar{\lambda}\langle u,v\rangle + \lambda\bar{\lambda}\langle v,v\rangle \\
&= \| u \|^2 - \lambda\overline{\langle u,v\rangle} - \bar{\lambda}\langle u,v\rangle + \lambda\bar{\lambda}\, \| v \|^2 \\
&= \| u \|^2 - \frac{|\langle u,v\rangle|^2}{\| v \|^2} - \frac{|\langle u,v\rangle|^2}{\| v \|^2} + \frac{|\langle u,v\rangle|^2}{\| v \|^2} \\
&= \| u \|^2 - \frac{|\langle u,v\rangle|^2}{\| v \|^2}.
\end{aligned}
$$

Therefore, $0 \le \| u \|^2 - \frac{|\langle u,v\rangle|^2}{\| v \|^2}$, or $|\langle u,v\rangle| \le \| u \| \; \| v \|$.

If the inequality holds as an equality, then $\| u - \lambda \cdot v \| = 0$, and so $u - \lambda \cdot v = 0$, thus u and v are linearly dependent. On the other hand, if u and v are linearly dependent, then $|\langle u,v\rangle| = \| u \| \; \| v \|$, as shown in the first proof.

More Proofs

There are indeed many different proofs of the Cauchy–Schwarz inequality other than the above two examples. When consulting other sources, there are often two sources of confusion. First, some authors define $\langle \cdot,\cdot \rangle$ to be linear in the second argument rather than the first. Second, some proofs are only valid when the field is R and not C.

Special Cases

R² (Ordinary Two-dimensional Space)

In the usual 2-dimensional space with the dot product, let $v = (v_1, v_2)$ and $u = (u_1, u_2)$. The Cauchy–Schwarz inequality is that

$$(\langle u,v\rangle)^2 = (\| u \| \; \| v \| \cos\theta)^2 \le \| u \|^2 \| v \|^2$$

where θ is the angle between u and v.

The form above is perhaps the easiest in which to understand the inequality, since the square of the cosine can be at most 1, which occurs when the vectors are in the same or opposite directions. It can also be restated in terms of the vector coordinates v_1, v_2, u_1 and u_2 as

$$(u_1 v_1 + u_2 v_2)^2 \le (u_1^2 + u_2^2) \cdot (v_1^2 + v_2^2)$$

where equality holds if and only if the vector (u_1, u_2) is in the same or opposite direction as the vector (v_1, v_2), or if one of them is the zero vector.

$\mathbb{R}^n$ (*n*-dimensional Euclidean Space)

In Euclidean space $\mathbb{R}^n$ with the standard inner product, the Cauchy–Schwarz inequality is

$$\left(\sum_{i=1}^{n} u_i v_i\right)^2 \leq \left(\sum_{i=1}^{n} u_i^2\right)\left(\sum_{i=1}^{n} v_i^2\right)$$

The Cauchy–Schwarz inequality can be proved using only ideas from elementary algebra in this case. Consider the following quadratic polynomial in x

$$0 \leq (u_1 x + v_1)^2 + \cdots + (u_n x + v_n)^2 = \left(\sum u_i^2\right)x^2 + 2\left(\sum u_i v_i\right)x + \sum v_i^2.$$

Since it is nonnegative, it has at most one real root for x, hence its discriminant is less than or equal to zero. That is,

$$\left(\sum (u_i v_i)\right)^2 - \sum u_i^2 \cdot \sum v_i^2 \leq 0,$$

which yields the Cauchy–Schwarz inequality.

L^2

For the inner product space of square-integrable complex-valued functions, one has

$$\left|\int_{\mathbb{R}^n} f(x)\overline{g(x)}dx\right|^2 \leq \int_{\mathbb{R}^n} |f(x)|^2 dx \cdot \int_{\mathbb{R}^n} |g(x)|^2 dx.$$

A generalization of this is the Hölder inequality.

Applications

Analysis

The triangle inequality for the standard norm is often shown as a consequence of the Cauchy–Schwarz inequality, as follows: given vectors x and y:

$$\begin{aligned}
\| x+y \|^2 &= \langle x+y, x+y \rangle \\
&= \| x \|^2 + \langle x, y \rangle + \langle y, x \rangle + \| y \|^2 \\
&= \| x \|^2 + 2\langle x, y \rangle + \| y \|^2 \\
&\leq \| x \|^2 + 2|\langle x, y \rangle| + \| y \|^2 \\
&\leq \| x \|^2 + 2\| x \| \, \| y \| + \| y \|^2 \\
&= (\| x \| + \| y \|)^2
\end{aligned}$$

Taking square roots gives the triangle inequality.

The Cauchy–Schwarz inequality is used to prove that the inner product is a continuous function with respect to the topology induced by the inner product itself.

Geometry

The Cauchy–Schwarz inequality allows one to extend the notion of "angle between two vectors" to any real inner product space, by defining:

$$\operatorname{Cos}\theta_{xy} = \frac{\langle x, y\rangle}{\|x\|\|y\|}.$$

The Cauchy–Schwarz inequality proves that this definition is sensible, by showing that the right-hand side lies in the interval [−1, 1], and justifies the notion that (real) Hilbert spaces are simply generalizations of the Euclidean space. It can also be used to define an angle in complex inner product spaces, by taking the absolute value or the real part of the right-hand side, as is done when extracting a metric from quantum fidelity.

Probability Theory

Let X, Y be random variables, then the covariance inequality is given by:

$$\operatorname{Var}(Y) \geq \frac{\operatorname{cov}(Y,X)\operatorname{cov}(Y,X)}{\operatorname{Var}(X)}.$$

After defining an inner product on the set of random variables using the expectation of their product,

$$\langle X, Y\rangle := E(XY),$$

then the Cauchy–Schwarz inequality becomes

$$|E(XY)|^2 \leq E(X^2)E(Y^2).$$

To prove the covariance inequality using the Cauchy–Schwarz inequality, let $\mu = E(X)$ and $\nu = E(Y)$, then

$$\begin{aligned}
&|\operatorname{Cov}(X,Y)|^2 = |E((X-\mu)(Y-\nu))|^2 \\
&= |\langle X-\mu, Y-\nu\rangle|^2 \\
&\leq \langle X-\mu, X-\mu\rangle\langle Y-\nu, Y-\nu\rangle \\
&= E((X-\mu)^2)E((Y-\nu)^2) \\
&= \operatorname{Var}(X)\operatorname{Var}(Y),
\end{aligned}$$

where Var denotes variance and Cov denotes covariance.

Generalizations

Various generalizations of the Cauchy–Schwarz inequality exist in the context of operator theory, e.g. for operator-convex functions, and operator algebras, where the domain and/or range are replaced by a C*-algebra or W*-algebra.

An inner product can be used to define a positive linear functional. For example, given a Hilbert space $L^2(m), m$ being a finite measure, the standard inner product gives rise to a positive functional ϕ by $\phi(g) = \langle g, 1 \rangle$. Conversely, every positive linear functional ϕ on $L^2(m)$ can be used to define an inner product $\langle f, g \rangle_\phi := \phi(g^* f)$ where g^* is the pointwise complex conjugate of g. In this language, the Cauchy–Schwarz inequality becomes

$$|\phi(g^* f)|^2 \leq \phi(f^* f)\phi(g^* g)$$

which extends verbatim to positive functionals on C*-algebras:

Theorem (Cauchy–Schwarz inequality for positive functionals on C*-algebras) If ϕ is a positive linear functional on a C*-algebra A, then for all $a, b \in A$, $|\phi(b^* a)|^2 \leq \phi(b^* b)\phi(a^* a)$.

The next two theorems are further examples in operator algebra.

Theorem (Kadison–Schwarz inequality, named after Richard Kadison) If ϕ is a unital positive map, then for every normal element a in its domain, we have $\phi(a^* a) \geq \phi(a^*)\phi(a)$ and $\phi(a^* a) \geq \phi(a)\phi(a^*)$.

This extends the fact $\varphi(a^* a) \cdot 1 \geq \varphi(a)^* \varphi(a) = |\varphi(a)|^2$, when φ is a linear functional. The case when a is self-adjoint, i.e. $a = a^*$, is sometimes known as Kadison's inequality.

Theorem (Modified Schwarz inequality for 2-positive maps). For a 2-positive map ϕ between C*-algebras, for all a, b in its domain,

$$\phi(a)^* \phi(a) \leq \| \phi(1) \| \phi(a^* a), \text{ and}$$

$$\| \phi(a^* b) \|^2 \leq \| \phi(a^* a) \| \cdot \| \phi(b^* b) \|.$$

Orthogonal and Orthonormal Basis

Let X be a vector space over $\mathbb{K}$, $\mathbb{K} = \mathbb{R}$ or $\mathbb{C}$. We note that $\mathbb{R} \subseteq \mathbb{C}$ and the conjugation. Let $X \times X \xrightarrow{s} \mathbb{K}$ be a sesquilinear form which is hermitian; that is, $s(x,u) = \overline{s(u,x)}$. Then $s(x,u) + \overline{s(x,u)} = s(x,u) + s(u,x) \in \mathbb{R}$ for each x, u ∈ X. If now s is positive i.e.

$Q(x) = s(x, x) \geq 0$ (note that $s(x,x) = \overline{s(x,x)}$ because s is hermitian so that $s(x, x) \in \mathbb{R}$), let us write $a - s(x, x)$, $b = s(u, u)$, $c = s(u, u)$ so that $a \geq 0$, $b \geq 0$, $c\bar{c} = |c|^2 \geq 0$. For any $r \in \mathbb{R}$, we have $0 \leq s\,(x - ur\,\bar{c}, x - ur\,\bar{c}) = s(x, x) - s(x, u)r\,\bar{c} - s(u, x)rc + s(u, u)r^2c\,\bar{c}$ $= a - 2c\,\bar{c}\,r + br^2c\,\bar{c} = |c|^2br^2 - 2|c|^2r + a$.

The quadratic expression $|c|^2br^2 - 2|c|^2r + a$ positive iff its discriminant $4|c|^4 - 4|c|^2ba \leq 0$ and thus we have, if $c \neq 0$, $|c|^2 \leq ba$. In case $c = 0$, this is true any way since a, b are positive real numbers. This is all situations we have $|c|^2 \leq ab$ i.e. $|s(x, u)|^2 \leq s(x, x)\, s(u, u)$. When s is a positive hermitian form, which is positive-definite in the sense that $s(x, x) = 0$ forces $x \in X$ to be $0 \in X$, we shall denote $s(x, u)$ by $(x \mid u)$ and call it an inner product on X. In this notation, we have just proved

$$\left|(x|u)\right|^2 \leq \|x\|^2\,\|u\|^2 \qquad (1)$$

This is known as the Cauchy-Schwarz inequality; note that we wrote $\|x\|^2$ for $s(x, x)$. The number is called the norm of $x \in X$.

A vector space X over $\mathbb{K}(\mathbb{K} = \mathbb{R}, \mathbb{C})$ is called the inner product space if we have equipped it with a preferred inner product. (From the theory of quadratic equations, we know that if $At^2 + Bt + C = A\,(t - \alpha)(t - \beta)$. Assuming without loss of generality $\alpha > \beta$, we have $(t - \alpha) > 0 \Rightarrow (t - \beta) > 0$ and $(t - \beta) < 0 \Rightarrow (t - \alpha) < 0$ so that $(t - \alpha)\,(t - \beta) > 0$ always holds and the expression $At^2 + Bt + c$ has the same sign as A unless t has a value lying between α and β in which case the expression has the sign opposite to that of A. If α, β are real with $\alpha = \beta$ then $At^2 + Bt + C = A\,(t - \alpha)^2$ has the same sign which A has. If α and β are complex then

$At^2 + Bt + C = A\left[\left(t + \dfrac{B}{2A}\right)^2 + \dfrac{4AC - B^2}{4A^2}\right]$ with $4AC - B^2 \geq 0$ since the roots are complex and we conclude that again the expression $At^2 + Bt + C$ has the same sign which A has.

To sum up: Suppose A, B, C real numbers. Then:

(i) $At^2 + Bt + C \geq 0$ for all $t \in \mathbb{R}$ iff $B^2 - 4AC \leq 0$, $A > 0$ we have used this here with $A = |c|^2b$, $B = -2|c|^2$, $C = a$

(ii) $At^2 + Bt + C \leq 0$ for all $t \in \mathbb{R}$ iff $B^2 - 4AC \leq 0$, $A < 0$)

We say $X \times X \xrightarrow{s} \mathbb{K}$ is sesquilinear iff $s(\lambda x, u + w\mu) = \bar{\lambda}\, s(x, u) + \bar{\lambda}\, s(x, w)\mu$ and $s(x + u, w\mu) = s(x, w) + s(u, w)\mu$ i.e. s is conjugate-linear in the first and linear in the second variable; 'sesqui' means 'one and a half'. This is the 'physicists's convention'; the 'mathematician's convention' is linear in the first and conjugate-linear in the second variable. Clearly, if s is sesquilinear in the physicist' convention, $\bar{s}$ given by $\bar{s}\,(x, u) := s(u, x)$ is sesquilinear in the mathematician's convention; one can adhere to either of the two. 'Conjugate-linear' is also called 'semi-linear' whence the name 'sesqui-linear'.

Suppose now X is an inner-product space. Then $(x|u) = (z|u)$ for each $u \in X$ means $(x - z|x - z) = (x|x) - (x|z) - (z|x) + (z|z) = (z|x) - (z|z) - (z|x) + (z|z) = 0$ and since $s(w, u) := (w|u)$ is positive-definite (so that $(w|w) = 0 \Rightarrow w = 0$), we get $x - z = 0$ i.e $x = z$. Similarly, $(u|x) = (u|z)$ for each $u \in X$ means $(x - z|x - z) = (x|x) - (x|z) - (z|x) + (z|z) = (x|z) - (x|z) - (z|z) + (z|z) = 0$ and again, since $s(w, u) := (w|u)$ is positive-definite, we get $x - z = 0$ i.e. $x = z$.

To sum up:

If s is positive -definite (and in particular if s is an inner product) $s(u,x) = s(u,z)$ for each $u \in X$ forces $x = z$ (in particular $(u|x) = (u|z)$ for each $u \in X$ forces $x = z$); similarly, $s(x,u) = s(z,u)$ for each $u \in X$ forces $x = z$). Further, $(x|u) = 0$ for all $u \in X$ means $(x|x) = 0$ as well and then positive-definiteness ensures $x = 0$.

Since the inner product is positive, $\|x\| = +\sqrt{(x|x)} \geq 0$; further, since it is positive-definite $\|x\| = +\sqrt{(x|x)} = 0$ forces $x = 0$. For $\lambda\lambda \in \mathbb{K}\,\mathbb{K}\,(\mathbb{K} = \mathbb{C} \text{ or } \mathbb{R})$ we have $\|x\lambda\| = +\sqrt{(x\lambda|x\lambda)} = \sqrt{\bar{\lambda}(x|x)\lambda} = \sqrt{\bar{\lambda}\lambda(x|x)} = |\lambda|\|x\| = \|x\||\lambda|$ ($\because$ $(x \mid x) \in \mathbb{R}$). Further, $\|x+u\|^2 = (x+u|x+u) = (x|x)+(x|u)+(u|x)+(u|u) = \|x\|^2 + 2(x|u) + \|u\|^2$

$\leq \|x\|^2 + 2|(x|u)| + \|u\|^2 \leq \|x\|^2 + 2\|x\|\|u\| + \|u\|^2 = (\|x\| + \|u\|)^2$. To sum up, we have proved (using the Cauchy-Schwarz inequality in the last argument).

The norm function $X \xrightarrow{x \to \|x\|} [0,\infty)$ satisfies:

(i) $\|x\| = 0 \Rightarrow x = 0$ (of course $\|x\| \geq 0$ and $\|0\| = 0$ hold)

(ii) $\|x\lambda\| = \|x\||\lambda| = |\lambda|\|x\|$

(iii) $\|x+u\| \leq \|x\| + \|u\|$

for all $x, u \in X, \lambda \in \mathbb{H}$ (or $\lambda \in \mathbb{C}$ or $\lambda \in \mathbb{R}$)

The inequality $\|x+u\| \leq \|x\| + \|u\|$ is called the triangle inequality.

In an inner product space, $\|u \pm x\|^2 = (u \pm x|u \pm x) = \|u\|^2 \pm 2(u|x) + \|x\|^2$ so that $\|u+x\|^2 + \|u-x\|^2 = [\|u\|^2 + \|x\|^2]$ for each $u, x \in X$. We refer to this equation as the parallelogram law.

We write $u \perp x$ (read: u is orthogonal to x) iff $(u|x) = 0$; this is clearly a symmetric relation ($u \perp x$ iff $x \perp u$).

The Cauchy-Schwarz inequality $|(u|x)| \le \|u\|\|x\|$ ensures that if u, x are nonzero, we have $\frac{|(u|x)_0|}{\|u\|\|x\|} \le \frac{|(u|x)|}{\|u\|\|x\|} \le 1$ so that $-1 \le \frac{(u|x)_0}{\|u\|\|x\|} \le 1$.

We define $\theta = \cos^{-1}\frac{(u|x)_0}{\|u\|\|x\|}$ to be the angle between u and x. This is called

(i) obtuse if $(u \mid x)_0 < 0$, and (ii) acute if $(u|x)_0$)0; we see it is a right angle if $(u|x)_0 = 0$. Note that now $\|u \pm x\|^2 = \|x\|^2 \pm 2\|x\|\|u\|\cos\theta + \|u\|^2$. Clearly, if $\cos\theta = 0$ we get $\|u \pm x\|^2 = \|x\|^2 + \|u\|^2$. Moreover, $u \perp x \Rightarrow \cos\theta = 0$ but $\cos\theta = 0 \Rightarrow u \perp x$ unless $\mathbb{K} = \mathbb{R}$.

A set $\{x_\alpha \in X\}$ is called an orthogonal set iff $x_\alpha \perp x_\beta$ whenever $\alpha \ne \beta$. Then $\left\{\frac{x_\alpha}{\|x_\alpha\|}\right\}$ also has this property if each $x_\alpha \ne 0$ and each vector in this set has norm 1.

(i) A set $\{x_\alpha \in X\}$ is called an orthonormal set iff $(x_\alpha|x_\beta) = \delta^\alpha_\beta$ (δ^α_β is the 'Kronecker delta', $\delta^\alpha_\beta = \begin{cases} 1 \text{ if } \alpha = \beta \\ 0 \text{ otherwise} \end{cases}$

(ii) An orthogonal set of nonzero vectors must be linearly independent

(If $\{x_\alpha\}$ is orthogonal and $x_1\lambda^1 + \cdots + x_n\lambda^n = 0$ then $0 = (x_1|0) = (x_i|x_1\lambda^1 + \cdots + x_n\lambda^n) = (x_i| x_1\lambda^1 + \cdots + x_n\lambda^n) = (x_i|x_1)\lambda^1 + \cdots + (x_i|x_n)\lambda^n = \|x_i\|\lambda^i$, ($\because (x_i|x_j) =$

$$\delta^i_j = \begin{cases} \|x_i\| \text{ if } i = j \\ 0 \text{ otherwise} \end{cases}$$

thus $\lambda^i = 0$ for $1 \le i \le n$ because $\|x\| \ne 0$ for any i).

(iii) Suppose $\{x_\alpha\}$ is orthonormal and $u \perp x_\alpha$ for each α forces u = 0. Then we can surely find no $u \in X$ with $\|u\| = 1$ such that $u \perp x_\alpha$ for each α which means that $\{x_\alpha\}$ cannot be enlarged to a bigger orthonormal set and if $\{x_\alpha\} \cup \{u\}$ is orthonormal, u must be one of the $\{x_\alpha\}$. Thus, if we decide to call a set total orthonormal set when it is an orthonormal set which cannot be enlarged to a bigger orthonormal set with the property that $x \in X$, $x \perp u_\alpha$ for each α forces x to be 0, $\{u_\alpha\}$ is total. A total orthonormal set will also be a linearly independent set (since every orthonormal set is linearly independent but it is not necessarily a maximal linearly independent set since it is not proved that it cannot be enlarged to a bigger linearly independent set; what is proved is that it cannot be enlarged to a bigger orthonormal set. Thus a total orthonormal set is not necessarily a Hamal basis of the vector space under consideration. Hamel basis = basis of a vector space defined earlier) However, we have the following

Theorem 0.1 If X is a finite dimensional innerproduct space over $\mathbb{K}(\mathbb{C} \text{ or } \mathbb{R})$ then every total orthonromal set is a Hamel basis of X.

Proof: Suppose $\{x_o, \cdots, x_{n-1}\}$ is an orthogonal set of non-zero vectors in an innerproduct space X having the property that $(x_i|x) = o$ for $o \le i \le n - 1$ forces $x = o$. Let $x \in X$ and $z = \sum_{i=0}^{n-1} x_i (x_i|x) = x_0 (x_0|x) + \cdots + x_{n-1}(x_{n-1}|x)$; then $(x_i|z) = (x_i|x)$ so that $(x_i|z - x) = 0$ for $o \le i \le n - 1$ and we have $x = z = \sum_{i=0}^{n-1} x_i (x_i|x)$ which shows that x is in the vector space generated by $\{x_o, \cdots, x_{n-1}\}$. On the other hand, $\{x_o, \cdots, x_{n-1}\}$ is linearly independent (since every orthogonal set of nonzero vectors is). Thus dim X = n and $\{x_o, \cdots, x_{n-1}\}$ is a Hamel basis of X.

In any innerproduct space X, one says that a sequence $\{x_n \in X\}$ is a Cauchy sequence iff it is possible to find an integer k such that $\|x_k - x_{n+k}\| \to 0$ as $n \to \infty$ and one says that the innerproduct space is a Hilbert space iff each Cauchy sequence is convergent in the sense that if $\{x_n \in X\}$ is a Cauchy sequence, one can find $x \in X$ with $\|x_n - x\| \to 0$ as $n \to \infty$; the vector $x \in X$ is then called the limit of {xn} and one writes $x_n \to x$ or $x = \lim_{n \leftarrow \infty} x_n$.

Each sequence $\{x_n\}$ yields a new sequence $s_n := \sum_{j=1}^{n} x_j$ and one writes $x = \sum_{j=1}^{\infty} x_j$ iff $s_n \to x$ saying that the series $\sum x_j$ converges to x.

Theorem o.2 When $\{x_n\}_{n=0}^{\infty}$ is a linearly independent set in an innerproduct space X, $y_n = x_n - \sum_{j=0}^{n-1} u_j (u_j|x_n); u_n = \frac{y_n}{\|y_n\|}, y_0 = x_0$ supply an orthonormal set $\{u_n\}_{n=0}^{\infty}$ such that $\text{span}\{u_o, \cdots, u_{n-1}\} = \text{span}\{x_o, \cdots, x_{n-1}\}$.

Proof: If $\{x_o\}$ is linearly indpendent, $y_0 = x_0 \neq 0$ and $u_0 = \frac{y_0}{\|y_0\|}$ is well defined and clearly $\text{span}\{u_o\} = \text{span}\{x_o\}$ with $\{u_o\}$ orthonormal. Assume now that $y_o, \cdots, y_{n-1}$ and $u_o, \cdots, u_{n-1}$ have been defined as above with $\{x_o, \cdots, x_{n-1}\}$ linearly independent and $\text{span}\{u_o, \cdots, u_{n-1}\} = \text{span}\{x_o, \cdots, x_{n-1}\}$, $u_o, \cdots, u_{n-1}$ orthonormal. Then $y_n = x_n - \sum_{j=0}^{n-1} (u_j|x_n) \neq 0$ if we have $\{x_o, \cdots, x_{n-1}, x_n\}$ linearly independent and $u_n = \frac{y_n}{\|y_n\|}$ is well defined. Further, for $o \le m \le n - 1$,

$$(u_m|u_n) = \left(u_m \middle| x_n - \sum_{j=0}^{n-1} u_j (u_j|x_n)\right) \frac{1}{\|y_n\|}$$

$$= [(u_m|x_n) - (u_m|u_m)(u_m|x_n)]\frac{1}{\|y_n\|}(\because (u_m|u_j) = \delta^j_m \text{ for } 0 \le j \le n-1)$$

$$= [(u_m|x_n) - (u_m|x_n)]\frac{1}{\|y_n\|}(\because (u_m|u_m) = 1)$$

$$= 0$$

Thus $u_n \perp u_m$ for $0 \le m \le n - 1$ and of course $(u_n|u_n) = \frac{1}{\|y_n\|}(y_n|y_n)\frac{1}{\|y_n\|} = 1$ so that $\{u_0, \cdots, u_n\}$ is an orthonormal set. Further, $x_n = u_n\|y_n\| + \sum_{j=0}^{n-1} u_j(u_j|x_n) = \sum_{j=1}^{n} u_j(u_j|x_n)$ $(\because (u_n|x_n) = (u_n|y_n) = \sum_{j=0}^{n-1}(u_n|u_j)(u_j|x_n) = (u_n|u_n|y_n\|) = \|y_n\|)$ and thus $x_n \in \text{span}\{u_0, \cdots, u_n\}$ proving that $\text{span}\{u_0, \cdots, u_n\} = \text{span}\{x_0, \cdots, x_n\}$ since it is known that $\text{span}\{x_0, \cdots, x_{n-1}\} = \text{span}\{u_0, \cdots, u_{n-1}\}$. Thus, in particular, if we have a finite dimensional Hilbert space, an orthonormal Hamel basis exists for it. The process of construction an orthonormal set out of a linearly independent set outlined in the theorem is called Gram-Schmidt orthonormalization.

Definition say a function $X \xrightarrow{T} X$ (X an innerproduct space) is adjointable iff there is a function $X \xrightarrow{T^+} X$ (read T + as "T dagger") such that the 'adjointness condition' $(T^+x|u) = (x|Tu)$ for each $x, u \in X$ is satisfied; we say T^+ is the Hilbert adjoint of T.

Proposition 0.3 If $X \xrightarrow{T} X$ is adjointable, the Hilbert adjoint T^+ is unique; further, both $X \xrightarrow{T} X$ and $X \xrightarrow{T^+} X$ are linear.

Proof:

(1) We first note that if $x = x'$, we have $(T^+x|u) = (x|Tu) = (x'|Tu) = (T^+x'|u)$ at each $u \in X$ so that (by positive-definiteness) we have $T^+x = T^+x'$. Thus the adjointness condition, if it holds, does define a function $X \xrightarrow{T^+} X$.

(2) If there are two Hilbert adjoints, say T^+ and $T^@$, for T, we have $(T^+x|u) = (x|Tu) = (T^@x|u)$ at each $u \in X$ and hence (by positive-definiteness) $T^+x = T^@x$; since this holds at each $x \in X$, we have $T^+ = T^@$. Thus there is at most one Hilbert adjoint for T.

(3) If the Hilbert adjoint T^+ exists, we have $(w|T(x + u\lambda)) = (T^+w|x + u\lambda) = (T^+w|x) + (T^+w|u)\lambda = (w|Tx) + (w|Tu)\lambda = (w|Tx + (Tu)\lambda)$ at each $w \in X$ so that (by positive-definiteness again) we have $T(x + u\lambda) = Tx + (Tu)\lambda$; this being true at each $x, u \in X, \lambda \in \mathbb{K}$, we see that T is (right) linear. Further, $(T^+(x + w\lambda)|u) = (x + w\lambda|Tu) = (x|Tu) + (w\lambda|Tu) = (x|Tu) + \bar{\lambda}(w|Tu) = (T^+x|u) + \bar{\lambda}(T^+w|u) = (T^+x|u) + ((T^+w)\lambda|u) = ((T^+x + (T^+w)\lambda)|u)$ for each $u \in X$.

Therefore (by positive-definiteness) we get

$T^{+} = (x + w\lambda) = T^{+}(x) + (T^{+}(W))\lambda$, and this being true at each $x, w \in X, \lambda \in \mathbb{K}$, we conclude that $X \xrightarrow{T^{+}} X$ is (right) linear.

The Next question is to determine which $X \xrightarrow{T} X$ are adjointable. The question makes sense clearly for linear operators $X \xrightarrow{T} X$ only.

Suppose X is finite dimensional, say dim X = n < ∞ coorinatizing by say $(\underline{\underline{e}}, \underline{\underline{\epsilon}}), \underline{\underline{\epsilon}}$ is the dual basis of $\underline{\underline{e}}$, we already know that $X \times X \xrightarrow{s} \mathbb{K}$ is a hermitian sesquilinear form iff $\left[s_i^j\right]_{n \times n} \left(s_I^j = \left(e_j, e_i\right)\right)$ is a hermitian matrix i.e. $s_i^j = \overline{s_j^i}$.

Thus s_j^i must be real (for $\mathbb{K} = \mathbb{C}$). For an inner product on X (thus X is the finite dimensional Hilbert space) we choose an orthonormal basis e (which is possible by Gram-Schmidt) and

note $(x|u) = (\sum_{i=1}^{n-1} e_i x^i | \sum_{j=0}^{n-1} e_j u^j) = \sum_{i=0}^{n-1}\sum_{j=0}^{n-1} \overline{x^i}(e_i|e_j)u^j = \sum_{i=0}^{n-1} \overline{x^i} u^i (\because (e_i|e_j) = 0 \text{ for } i \neq j \; (e_i|e_i) = 1)$.

Thus $(x|u) = x^* u$ where x^* is the conjugate-transpose of the column vector $x = \begin{pmatrix} x^0 \\ . \\ . \\ . \\ x^{n-1} \end{pmatrix}_{n \times 1}$ and

u is the column vector $\begin{pmatrix} u^0 \\ . \\ . \\ . \\ u^{n-1} \end{pmatrix}_{n \times 1}$ (being calculated with respect to (e, ϵ)).

Therefore $(u \mid Tx) = u^* Tx = u^*(x^* T^*)^*$ (T being the matrix of T with respect to (e, ϵ) and T^* its conjugate-transpose)

$= (x^* T^* u)^* = (T^* u)^* x \; (\because x^{**} = x)$

$= (T^{@} u \mid x)$

Where $X \xrightarrow{T^{@}} X$ is the operator having the matrix T^* with respect to (e, ϵ) and sending u to $T^* u$. But since $(u|Tx) = (T^{+}u|x)$ and the Hilbert adjoint T^{+} is unique, we conclude that $T^{@} = T^{+}$ and thus the matrix of T^{+} with respect to (e, ϵ) T^*.

To sum up:

Every linear operator $X \xrightarrow{T} X$ is adjointable when X is a finite dimensional Hilbert space. This result does not hold in infinite dimensional inner product space in general (when X is a Hilbert space and $\|T\| := \sup_{\|x\|\le 1}\|Tx\| < \infty$, it is however true. We are not concerned with these considerations in this course).

Now take $\mathbb{K} = \mathbb{C}$s that X is a finite dimensional complex Hilbert space, say $\mathbb{C}^n$. Let L(X, X) be the space of all linear operators X → X. Then we know that L(X, X) has dimension n^2 and if (e, ϵ) is a coordinatization of $\mathbb{C}^2$,

$\left\{ |e_j\rangle\langle\epsilon^i| \,\middle|\, \begin{matrix} 0 \le i \le n-1 \\ 0 \le j \le n-1 \end{matrix} \right\}$ is a basis of L(X, X).

For $X \xrightarrow{A} X$, we define $\mathrm{trace}(A) := \sum_{i=0}^{n-1} \langle \epsilon^i | Ae_i \rangle$; (Thus if we denote the associated matrix of

A by $a = [a_i^j]$ we have $\mathrm{trace}(A) := \sum_{i=0}^{n-1} a_i^i$; we then define the trace of a square matrix to be the sum of its diagonal elements and in view of the bijectivity $\mathbb{K}_{n\times n} \overset{A\leftrightarrow a}{\leftrightarrow} \mathrm{Lim}(X, X)$ (just define A by Ax := ax for x ∈ X) the two definitions record the same mathematical concept). Clearly, $L(X, X) \xrightarrow{\mathrm{trace}} \mathbb{C}$ is a linear form.

If (d, δ) δ is the dual basis of d, any other coordinatization for X we use the decomposition of identity $X \xrightarrow{id} X = X \xrightarrow{\sum |e_i\rangle\langle\epsilon^i|} X = X \xrightarrow{\sum |d_j\rangle\langle\delta^j|} X$ to find

$$\mathrm{trace}(A) := \sum_{i=0}^{n-1} \langle \epsilon^i | Ae_i \rangle = \sum_{i=0}^{n-1} \langle \epsilon^i | A(|de_i) \rangle$$

$$= \sum_{i=0}^{n-1}\sum_{j=0}^{n-1} \langle \epsilon^i | Ad_j \rangle \langle \delta^j | e_i \rangle$$

$$= \sum_{j=0}^{n-1}\sum_{i=0}^{n-1} \langle \delta^j | e_i \rangle \langle \epsilon^i | Ad_j \rangle$$

$$= \sum_{j=0}^{n-1} \langle \delta^j || d(Ad_j) \rangle = \sum_{j=0}^{n-1} \langle \delta^j | Ad_j \rangle$$

which means that the trace of A is coordinate-free (trace (A) is the same whatever coordintization is chosen).

Considering $(X(e,\epsilon)) \xrightarrow[a,b]{A,B} (X(e,\epsilon))$ we define (A|B):= trace (A^+B) = trace (a^*b) ∈ $\mathbb{C}$ (since a^* is the matrix of A^+ as noted above). Then

(i) (λA + B | C + Dμ) = trace ((λa + b)*(c + dμ)) = trace ((b^* + $a^*\bar{\lambda}$)(c + dμ))

= trace ($\bar{\lambda}$ a^*c + b^*c + $\bar{\lambda}$ $a^*d\mu$ + $b^*d\mu$)

= $\bar{\lambda}$ trace (a^*c) + trace (b^*c) + $\bar{\lambda}$ trace (a^+d)μ + trace (b^*d)μ

= $\bar{\lambda}$ (A | C) + (B | C) + $\bar{\lambda}$ (A | D)μ + (B | D)μ which says that (A|B) as defined is a sesquilinear form on L(X, X).

(ii) (A|B) = trace (a*b) = $\overline{\text{trace}\left(b^*\left(a^*\right)^*\right)} = \overline{\text{trace}\left(b^*a\right)} = \overline{\left(B|A\right)}$ which says it is hermitian, and

(iii) (A|A) = trace (a*a) = $\sum_{i=0}^{n-1}\left(a^*a\right)_i^i = \sum_{i=0}^{n-1}\sum_{j=0}^{n-1}\left(a^*\right)_j^i a_i^j$

$= \sum_{i=0}^{n-1}\sum_{j=0}^{n-1}\overline{a_i^j}a_i^j = \sum_{i=0}^{n-1}\sum_{j=0}^{n-1}\left|a_i^j\right|^2 =$ Which says (A |A) ≥ 0, (A|A) = 0 iff each $a_i^j = 0, 0 \le i \le n-1,$ $0 \le i \le j - 1$ i.e. $a = 0$ i.e. A = 0 i.e. this is positivedefinite.

Thus (A, B) → (A|B) = trace (A⁺B) = trace (a*b) defines an inner product on L(X, X) into a complex Hilbert space of dimension n^2. This inner product is called the Hilbert-Schmidt inner product on L(X, X) and equipped with this inner product, L(X, X) is called the Liouville space of the Hilbert space X. (the name 'Liouville space' is used mostly in quantum mechanics).

(iv) If dim V = n (∞, we have essentially V = $\mathbb{C}^n$ and V*, the dual, also has dimension n when V is an ips, let us write, given z ∈ V, $V \xrightarrow{\varphi_z} \mathbb{C}$ supplied by $\phi_z(x) := (z|x)$. Then $\phi_z(x + u\lambda) = (z|x + u\lambda) = (z|x) + (z|u)\lambda = \phi_z(x) + (\phi_z(u))\lambda$ so that $\phi_z \in V^{tr}$ Further, $V \xrightarrow{\varphi} V^{tr}$ defined by $\phi(z) := \phi_z$ obeys $(\phi(\lambda z + w))(x) = (\lambda z + w \mid x) = \bar{\lambda}(z|x) + (w|x) = \bar{\lambda}\phi_z(x) + \phi_w(x)$ at each x ∈ X which means $\phi(\lambda z + w) = \bar{\lambda}\phi(z) + \phi(w)$ showing that $V \xrightarrow{\varphi} V^{tr}$ is conjugate-linear. If $\phi_z = \phi_w$ then at each x ∈ V we have $\phi_z(x) = (z|x) = \phi_w(x) = (w|x)$ which means (by positive definiteness) z = w. Thus this ϕ is a bijective (injectivity would ensure this because dim V = dim V^{tr}(∞). Explicitly, Suppose f ∈ V^{tr} and choose an orthonormal basis $\{e_0, \cdots, c_{n-1}\}$ of V ; with respect to this basis of V (and the basis {1} of $\mathbb{C}$ the linear form $V \xrightarrow{f} \mathbb{C}$ will be given by a row vector $[a_0, \cdots, a_{n-1}] \in \mathrm{Mat}_{1\times n}\mathbb{C}$ so that for any $x = \sum_{i=0}^{n-1} e_i x^i$ we have $f(x) = \sum_{i=0}^{n-1} f(e_i)x^i, f(e_i) = a_i$ so that if $a = \sum \overline{a_j} e_j$, we have $\phi_a(x) = (a|x) = \left(\sum \overline{a_j} e_j \middle| \sum e_i x^i\right) = \sum a_i x^i = f(x)$ at each x ∈ V and thus f = ϕ_a.

To sum up:

Riesz Representation Theorem: If dim V ⟨∞, there exists a conjugate-linear bijection $V \xrightarrow{T} V^{tr}$ given by T(z) ∈ V^{tr} computing as (T (z))(x) = (z | x) for z, x ∈ V. Given f ∈ V^{tr}, it is

$$z = \begin{pmatrix} \overline{a_0} \\ \cdot \\ \cdot \\ \cdot \\ \overline{a_{n-1}} \end{pmatrix} \in V$$ which is such that T (z) = f; we say z is the representor for f and this column vector is with reference to a chosen orthonormal basis $\underline{\underline{e}}$; $a_i = f(e_i)$.

Orthonormality

In linear algebra, two vectors in an inner product space are orthonormal if they are orthogonal and unit vectors. A set of vectors form an orthonormal set if all vectors in the set are mutually orthogonal and all of unit length. An orthonormal set which forms a basis is called an orthonormal basis.

Intuitive Overview

The construction of orthogonality of vectors is motivated by a desire to extend the intuitive notion of perpendicular vectors to higher-dimensional spaces. In the Cartesian plane, two vectors are said to be *perpendicular* if the angle between them is 90° (i.e. if they form a right angle). This definition can be formalized in Cartesian space by defining the dot product and specifying that two vectors in the plane are orthogonal if their dot product is zero.

Similarly, the construction of the norm of a vector is motivated by a desire to extend the intuitive notion of the length of a vector to higher-dimensional spaces. In Cartesian space, the *norm* of a vector is the square root of the vector dotted with itself. That is,

$$\| x \| = \sqrt{x \cdot x}$$

Many important results in linear algebra deal with collections of two or more orthogonal vectors. But often, it is easier to deal with vectors of unit length. That is, it often simplifies things to only consider vectors whose norm equals 1. The notion of restricting orthogonal pairs of vectors to only those of unit length is important enough to be given a special name. Two vectors which are orthogonal and of length 1 are said to be *orthonormal.*

Simple Example

What does a pair of orthonormal vectors in 2-D Euclidean space look like?

Let $u = (x_1, y_1)$ and $v = (x_2, y_2)$. Consider the restrictions on x_1, x_2, y_1, y_2 required to make u and v form an orthonormal pair.

- From the orthogonality restriction, $u \bullet v = 0$.
- From the unit length restriction on u, $||u|| = 1$.
- From the unit length restriction on v, $||v|| = 1$.

Expanding these terms gives 3 equations:

1. $x_1 x_2 + y_1 y_2 = 0$
2. $\sqrt{{x_1}^2 + {y_1}^2} = 1$

3. $\sqrt{x_2^2 + y_2^2} = 1$

Converting from Cartesian to polar coordinates, and considering Equation (2) and Equation (3) immediately gives the result $r_1 = r_2 = 1$. In other words, requiring the vectors be of unit length restricts the vectors to lie on the unit circle.

After substitution, Equation (1) becomes $\cos\theta_1 \cos\theta_2 + \sin\theta_1 \sin\theta_2 = 0$. Rearranging gives $\tan\theta_1 = -\cot\theta_2$. Using a trigonometric identity to convert the cotangent term gives

$$\tan(\theta_1) = \tan\left(\theta_2 + \tfrac{\pi}{2}\right)$$

$$\Rightarrow \theta_1 = \theta_2 + \tfrac{\pi}{2}$$

It is clear that in the plane, orthonormal vectors are simply radii of the unit circle whose difference in angles equals 90°.

Definition

Let $\mathcal{V}$ be an inner-product space. A set of vectors

$$\{u_1, u_2, \ldots, u_n, \ldots\} \in \mathcal{V}$$

is called orthonormal if and only if

$$\forall i, j : \langle u_i, u_j \rangle = \delta_{ij}$$

where δ_{ij} is the Kronecker delta and $\langle \cdot, \cdot \rangle$ is the inner product defined over $\mathcal{V}$.

Significance

Orthonormal sets are not especially significant on their own. However, they display certain features that make them fundamental in exploring the notion of diagonalizability of certain operators on vector spaces.

Properties

Orthonormal sets have certain very appealing properties, which make them particularly easy to work with.

- Theorem. If $\{e_1, e_2, \ldots, e_n\}$ is an orthonormal list of vectors, then

$$\| a_1 e_1 + a_2 e_2 + \cdots + a_n e_n \|^2 = | a_1 |^2 + | a_2 |^2 + \cdots + | a_n |^2$$

- Theorem. Every orthonormal list of vectors is linearly independent.

Existence

- Gram-Schmidt theorem. If $\{v_1, v_2,...,v_n\}$ is a linearly independent list of vectors in an inner-product space $\mathcal{V}$, then there exists an orthonormal list $\{e_1, e_2,...,e_n\}$ of vectors in $\mathcal{V}$ such that span$(e_1, e_2,...,e_n)$ = span$(v_1, v_2,...,v_n)$.

Proof of the Gram-Schmidt theorem is constructive, and discussed at length elsewhere. The Gram-Schmidt theorem, together with the axiom of choice, guarantees that every vector space admits an orthonormal basis. This is possibly the most significant use of orthonormality, as this fact permits operators on inner-product spaces to be discussed in terms of their action on the space's orthonormal basis vectors. What results is a deep relationship between the diagonalizability of an operator and how it acts on the orthonormal basis vectors. This relationship is characterized by the Spectral Theorem.

Examples

Standard Basis

The standard basis for the coordinate space F^n is

$$\{e_1, e_2,...,e_n\} \text{ where } \begin{aligned} e_1 &= (1, 0, ..., 0) \\ e_2 &= (0, 1, ..., 0) \\ &\vdots \\ e_n &= (0, 0, ..., 1) \end{aligned}$$

Any two vectors e_i, e_j where $i \neq j$ are orthogonal, and all vectors are clearly of unit length. So $\{e_1, e_2,...,e_n\}$ forms an orthonormal basis.

Real-valued Functions

When referring to real-valued functions, usually the L^2 inner product is assumed unless otherwise stated. Two functions $\phi(x)$ and $\psi(x)$ are orthonormal over the interval $[a,b]$ if

$$(1) \quad \langle \phi(x), \psi(x) \rangle = \int_a^b \phi(x)\psi(x)dx = 0, \quad \text{and}$$

$$(2) \quad \| \phi(x) \|_2 = \| \psi(x) \|_2 = \left[\int_a^b |\phi(x)|^2 \, dx \right]^{\frac{1}{2}} = \left[\int_a^b |\psi(x)|^2 \, dx \right]^{\frac{1}{2}} = 1.$$

Fourier Series

The Fourier series is a method of expressing a periodic function in terms of sinusoidal basis functions. Taking $C[-\pi,\pi]$ to be the space of all real-valued functions continuous on the interval $[-\pi,\pi]$ and taking the inner product to be

$$\langle f, g \rangle = \int_{-\pi}^{\pi} f(x)g(x)dx$$

It can be shown that

$$\left\{ \frac{1}{\sqrt{2\pi}}, \frac{\sin(x)}{\sqrt{\pi}}, \frac{\sin(2x)}{\sqrt{\pi}}, \ldots, \frac{\sin(nx)}{\sqrt{\pi}}, \frac{\cos(x)}{\sqrt{\pi}}, \frac{\cos(2x)}{\sqrt{\pi}}, \ldots, \frac{\cos(nx)}{\sqrt{\pi}} \right\}, \quad n \in \mathbb{N}$$

forms an orthonormal set.

However, this is of little consequence, because $C[-\pi,\pi]$ is infinite-dimensional, and a finite set of vectors cannot span it. But, removing the restriction that n be finite makes the set dense in $C[-\pi,\pi]$ and therefore an orthonormal basis of $C[-\pi,\pi]$.

References

- P. K. Jain, Khalil Ahmad (1995). "5.1 Definitions and basic properties of inner product spaces and Hilbert spaces". Functional analysis (2nd ed.). New Age International. p. 203. ISBN 81-224-0801-X
- Aldaz, J.M.; Barza, S.; Fujii, M.; Moslehian, M.S. (2015), "Advances in Operator Cauchy—Schwarz inequalities and their reverses", Annals of Functional Analysis, 6 (3): 275–295, doi:10.15352/afa/06-3-20
- Emch, Gerard G. (1972). Algebraic methods in statistical mechanics and quantum field theory. New York: Wiley-Interscience. ISBN 978-0-471-23900-0
- Grinshpan, A. Z. (2005), "General inequalities, consequences, and applications", Advances in Applied Mathematics, 34 (1): 71–100, doi:10.1016/j.aam.2004.05.001
- Dragomir, S. S. (2003), "A survey on Cauchy–Bunyakovsky–Schwarz type discrete inequalities", Journal of Inequalities in Pure and Applied Mathematics, 4 (3): 142 pp
- Bityutskov, V.I. (2001), "Bunyakovskii inequality", in Hazewinkel, Michiel, Encyclopedia of Mathematics, Springer, ISBN 978-1-55608-010-4
- Kadison, R.V. (1952), "A generalized Schwarz inequality and algebraic invariants for operator algebras", Annals of Mathematics, 56 (3): 494–503, JSTOR 1969657, doi:10.2307/1969657
- Solomentsev, E.D. (2001), "Cauchy inequality", in Hazewinkel, Michiel, Encyclopedia of Mathematics, Springer, ISBN 978-1-55608-010-4

3

Understanding Matrices

A matrix can be defined as an array of numbers, symbols, etc. that are arranged in rows and columns. Matrices can be applied in a number of fields like classical mechanics, quantum mechanics, electromagnetism and quantum electromagnetism. The aspects elucidated in this chapter are of vital importance, and provide a better understanding of matrices.

Matrix (Mathematics)

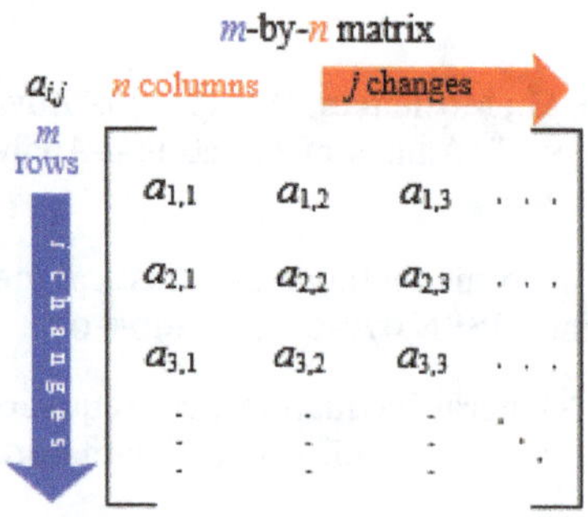

The m rows are horizontal and the n columns are vertical. Each element of a matrix is often denoted by a variable with two subscripts. For example, $a_{2,1}$ represents the element at the second row and first column of a matrix A.

In mathematics, a matrix (plural matrices) is a rectangular array of numbers, symbols, or expressions, arranged in rows and columns. For example, the dimensions of the matrix below are 2 × 3 (read "two by three"), because there are two rows and three columns:

$$\begin{bmatrix} 1 & 9 & -13 \\ 20 & 5 & -6 \end{bmatrix}.$$

The individual items in an $m \times n$ matrix A, often denoted by $a_{i,j}$, where max $i = m$ and max $j = n$, are called its *elements* or *entries*. Provided that they have the same size (each matrix has the same number of rows and the same number of columns as the other), two matrices can be added or subtracted element by element. The rule for matrix multiplication, however, is that two matrices can be multiplied only when the number of columns in the first equals the number of rows in the second (i.e., the inner dimensions are the same, n for $A_{m,n} \times B_{n,p}$). Any matrix can be multiplied element-wise by a scalar from its associated field. A major application of matrices is to represent linear transformations, that is, generalizations of linear functions such as $f(x) = 4x$. For example, the rotation of vectors in three-dimensional space is a linear transformation, which can be represented by a rotation matrix R: if **v** is a column vector (a matrix with only one column) describing the position of a point in space, the product Rv is a column vector describing the position of that point after a rotation. The product of two transformation matrices is a matrix that represents the composition of two linear transformations. Another application of matrices is in the solution of systems of linear equations. If the matrix is square, it is possible to deduce some of its properties by computing its determinant. For example, a square matrix has an inverse if and only if its determinant is not zero. Insight into the geometry of a linear transformation is obtainable (along with other information) from the matrix's eigenvalues and eigenvectors.

Applications of matrices are found in most scientific fields. In every branch of physics, including classical mechanics, optics, electromagnetism, quantum mechanics, and quantum electrodynamics, they are used to study physical phenomena, such as the motion of rigid bodies. In computer graphics, they are used to manipulate 3D models and project them onto a 2-dimensional screen. In probability theory and statistics, stochastic matrices are used to describe sets of probabilities; for instance, they are used within the PageRank algorithm that ranks the pages in a Google search. Matrix calculus generalizes classical analytical notions such as derivatives and exponentials to higher dimensions. Matrices are used in economics to describe systems of economic relationships.

A major branch of numerical analysis is devoted to the development of efficient algorithms for matrix computations, a subject that is centuries old and is today an expanding area of research. Matrix decomposition methods simplify computations, both theoretically and practically. Algorithms that are tailored to particular matrix structures, such as sparse matrices and near-diagonal matrices, expedite computations in finite element method and other computations. Infinite matrices occur in planetary theory and in atomic theory. A simple example of an infinite matrix is the matrix representing the derivative operator, which acts on the Taylor series of a function.

Definition

A *matrix* is a rectangular array of numbers or other mathematical objects for which operations such as addition and multiplication are defined. Most commonly, a matrix over a field *F* is a rectangular array of scalars each of which is a member of *F*. Most of this article focuses on *real* and *complex matrices*, that is, matrices whose elements are real numbers or complex numbers, respectively. More general types of entries are discussed below. For instance, this is a real matrix:

$$A = \begin{bmatrix} -1.3 & 0.6 \\ 20.4 & 5.5 \\ 9.7 & -6.2 \end{bmatrix}.$$

The numbers, symbols or expressions in the matrix are called its *entries* or its *elements*. The horizontal and vertical lines of entries in a matrix are called *rows* and *columns*, respectively.

Size

The size of a matrix is defined by the number of rows and columns that it contains. A matrix with m rows and n columns is called an $m \times n$ matrix or m-by-n matrix, while m and n are called its *dimensions*. For example, the matrix **A** above is a 3×2 matrix.

Matrices which have a single row are called *row vectors*, and those which have a single column are called *column vectors*. A matrix which has the same number of rows and columns is called a *square matrix*. A matrix with an infinite number of rows or columns (or both) is called an *infinite matrix*. In some contexts, such as computer algebra programs, it is useful to consider a matrix with no rows or no columns, called an *empty matrix*.

Name	Size	Example	Description
Row vector	$1 \times n$	$\begin{bmatrix} 3 & 7 & 2 \end{bmatrix}$	A matrix with one row, sometimes used to represent a vector
Column vector	$n \times 1$	$\begin{bmatrix} 4 \\ 1 \\ 8 \end{bmatrix}$	A matrix with one column, sometimes used to represent a vector
Square matrix	$n \times n$	$\begin{bmatrix} 9 & 13 & 5 \\ 1 & 11 & 7 \\ 2 & 6 & 3 \end{bmatrix}$	A matrix with the same number of rows and columns, sometimes used to represent a linear transformation from a vector space to itself, such as reflection, rotation, or shearing.

Notation

Matrices are commonly written in box brackets or parentheses:

$$\mathbf{A} = \begin{bmatrix} a_{11} & a_{12} & \cdots & a_{1n} \\ a_{21} & a_{22} & \cdots & a_{2n} \\ \vdots & \vdots & \ddots & \vdots \\ a_{m1} & a_{m2} & \cdots & a_{mn} \end{bmatrix} = \begin{pmatrix} a_{11} & a_{12} & \cdots & a_{1n} \\ a_{21} & a_{22} & \cdots & a_{2n} \\ \vdots & \vdots & \ddots & \vdots \\ a_{m1} & a_{m2} & \cdots & a_{mn} \end{pmatrix} = \left(a_{ij}\right) \in \mathbb{R}^{m\times n}.$$

The specifics of symbolic matrix notation vary widely, with some prevailing trends. Matrices are usually symbolized using upper-case letters (such as **A** in the examples above), while the corresponding lower-case letters, with two subscript indices (for example, a_{11}, or $a_{1,1}$), represent the entries. In addition to using upper-case letters to symbolize matrices, many authors use a special typographical style, commonly boldface upright (non-italic), to further distinguish matrices from other mathematical objects. An alternative notation involves the use of a double-underline with the variable name, with or without boldface style, (for example, $\underline{\underline{A}}$).

The entry in the *i*-th row and *j*-th column of a matrix **A** is sometimes referred to as the *i*,*j*, (*i*,*j*), or $(i,j)^{th}$ entry of the matrix, and most commonly denoted as $a_{i,j}$, or a_{ij}. Alternative notations for that entry are $A[i,j]$ or $A_{i,j}$. For example, the (1,3) entry of the following matrix **A** is 5 (also denoted a_{13}, $a_{1,3}$, $A[1,3]$ or $A_{1,3}$):

$$A = \begin{bmatrix} 4 & -7 & 5 & 0 \\ -2 & 0 & 11 & 8 \\ 19 & 1 & -3 & 12 \end{bmatrix}$$

Sometimes, the entries of a matrix can be defined by a formula such as $a_{i,j} = f(i, j)$. For example, each of the entries of the following matrix **A** is determined by $a_{ij} = i - j$.

$$A = \begin{bmatrix} 0 & -1 & -2 & -3 \\ 1 & 0 & -1 & -2 \\ 2 & 1 & 0 & -1 \end{bmatrix}$$

In this case, the matrix itself is sometimes defined by that formula, within square brackets or double parentheses. For example, the matrix above is defined as **A** = $[i\text{-}j]$, or **A** = $((i\text{-}j))$. If matrix size is $m \times n$, the above-mentioned formula $f(i, j)$ is valid for any $i = 1, \ldots, m$ and any $j = 1, \ldots, n$. This can be either specified separately, or using $m \times n$ as a subscript. For instance, the matrix **A** above is 3×4 and can be defined as **A** = $[i - j]$ ($i = 1, 2, 3; j = 1, \ldots, 4$), or **A** = $[i - j]_{3\times 4}$.

Some programming languages utilize doubly subscripted arrays (or arrays of arrays) to represent an m-×-n matrix. Some programming languages start the numbering of array indexes at zero, in which case the entries of an m-by-n matrix are indexed by $0 \le i \le m - 1$ and $0 \le j \le n - 1$. This article follows the more common convention in mathematical writing where enumeration starts from 1.

An asterisk is occasionally used to refer to whole rows or columns in a matrix. For example, $a_{i,*}$ refers to the ith row of **A**, and $a_{*,j}$ refers to the jth column of **A**. The set of all m-by-n matrices is denoted $\mathbb{M}(m, n)$.

Basic Operations

There are a number of basic operations that can be applied to modify matrices, called *matrix addition*, *scalar multiplication*, *transposition*, *matrix multiplication*, *row operations*, and *submatrix*.

Addition, Scalar Multiplication and Transposition

Operation	Definition	Example
Addition	The sum A+B of two m-by-n matrices A and B is calculated entry-wise: $(A + B)_{i,j} = A_{i,j} + B_{i,j}$, where $1 \le i \le m$ and $1 \le j \le n$.	$\begin{bmatrix}1 & 3 & 1\\ 1 & 0 & 0\end{bmatrix}+\begin{bmatrix}0 & 0 & 5\\ 7 & 5 & 0\end{bmatrix}=\begin{bmatrix}1+0 & 3+0 & 1+5\\ 1+7 & 0+5 & 0+0\end{bmatrix}=\begin{bmatrix}1 & 3 & 6\\ 8 & 5 & 0\end{bmatrix}$
Scalar multiplication	The product c**A** of a number c (also called a scalar in the parlance of abstract algebra) and a matrix **A** is computed by multiplying every entry of **A** by c: $(c\mathbf{A})_{i,j} = c \cdot \mathbf{A}_{i,j}$. This operation is called *scalar multiplication*, but its result is not named "scalar product" to avoid confusion, since "scalar product" is sometimes used as a synonym for "inner product".	$2\cdot\begin{bmatrix}1 & 8 & -3\\ 4 & -2 & 5\end{bmatrix}=\begin{bmatrix}2\cdot 1 & 2\cdot 8 & 2\cdot -3\\ 2\cdot 4 & 2\cdot -2 & 2\cdot 5\end{bmatrix}=\begin{bmatrix}2 & 16 & -6\\ 8 & -4 & 10\end{bmatrix}$
Transposition	The *transpose* of an *m*-by-*n* matrix **A** is the *n*-by-*m* matrix $\mathbf{A}^T$ (also denoted $\mathbf{A}^{tr}$ or ${}^t\mathbf{A}$) formed by turning rows into columns and vice versa: $(\mathbf{A}^T)_{i,j} = \mathbf{A}_{j,i}$.	$\begin{bmatrix}1 & 2 & 3\\ 0 & -6 & 7\end{bmatrix}^T=\begin{bmatrix}1 & 0\\ 2 & -6\\ 3 & 7\end{bmatrix}$

Familiar properties of numbers extend to these operations of matrices: for example, addition is commutative, that is, the matrix sum does not depend on the order of the summands: A + B = B + A. The transpose is compatible with addition and scalar multiplication, as expressed by $(cA)^T = c(A^T)$ and $(A + B)^T = A^T + B^T$. Finally, $(A^T)^T = A$.

Matrix Multiplication

Multiplication of two matrices is defined if and only if the number of columns of the left matrix is the same as the number of rows of the right matrix. If **A** is an *m*-by-*n* matrix and B is an n-by-p matrix, then their matrix product AB is the m-by-p matrix whose entries are given by dot product of the corresponding row of A and the corresponding column of B:

$$[AB]_{i,j} = A_{i,1}B_{1,j} + A_{i,2}B_{2,j} + \cdots + A_{i,n}B_{n,j} = \sum_{r=1}^{n} A_{i,r}B_{r,j},$$

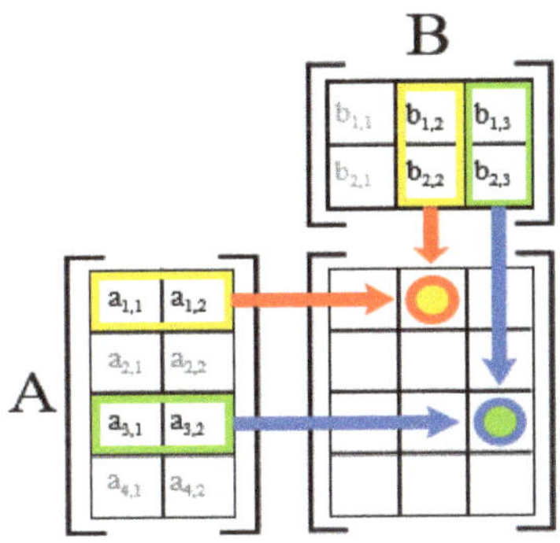

Schematic depiction of the matrix product AB of two matrices A and B.

where $1 \le i \le m$ and $1 \le j \le p$. For example, the underlined entry 2340 in the product is calculated as (2 × 1000) + (3 × 100) + (4 × 10) = 2340:

$$\begin{bmatrix} \underline{2} & \underline{3} & \underline{4} \\ 1 & 0 & 0 \end{bmatrix} \begin{bmatrix} 0 & \underline{1000} \\ 1 & \underline{100} \\ 0 & \underline{10} \end{bmatrix} = \begin{bmatrix} 3 & \underline{2340} \\ 0 & 1000 \end{bmatrix}.$$

Matrix multiplication satisfies the rules (AB)C = A(BC) (associativity), and (A+B)C = AC+BC as well as C(A+B) = CA+CB (left and right distributivity), whenever the size of the matrices is such that the various products are defined. The product AB may be defined without BA being defined, namely if A and B are m-by-n and n-by-k matrices, respectively, and $m \neq k$. Even if both products are defined, they need not be equal, that is, generally

AB ≠ BA,

that is, matrix multiplication is not commutative, in marked contrast to (rational, real, or complex) numbers whose product is independent of the order of the factors. An example of two matrices not commuting with each other is:

$$\begin{bmatrix} 1 & 2 \\ 3 & 4 \end{bmatrix} \begin{bmatrix} 0 & 1 \\ 0 & 0 \end{bmatrix} = \begin{bmatrix} 0 & 1 \\ 0 & 3 \end{bmatrix},$$

whereas

$$\begin{bmatrix} 0 & 1 \\ 0 & 0 \end{bmatrix} \begin{bmatrix} 1 & 2 \\ 3 & 4 \end{bmatrix} = \begin{bmatrix} 3 & 4 \\ 0 & 0 \end{bmatrix}.$$

Besides the ordinary matrix multiplication just described, there exist other less frequently used operations on matrices that can be considered forms of multiplication, such as the Hadamard product and the Kronecker product. They arise in solving matrix equations such as the Sylvester equation.

Row Operations

There are three types of row operations:

1. row addition, that is adding a row to another.
2. row multiplication, that is multiplying all entries of a row by a non-zero constant;
3. row switching, that is interchanging two rows of a matrix;

These operations are used in a number of ways, including solving linear equations and finding matrix inverses.

Submatrix

A submatrix of a matrix is obtained by deleting any collection of rows and/or columns. For example, from the following 3-by-4 matrix, we can construct a 2-by-3 submatrix by removing row 3 and column 2:

$$A = \begin{bmatrix} 1 & 2 & 3 & 4 \\ 5 & 6 & 7 & 8 \\ 9 & 10 & 11 & 12 \end{bmatrix} \rightarrow \begin{bmatrix} 1 & 3 & 4 \\ 5 & 7 & 8 \end{bmatrix}.$$

The minors and cofactors of a matrix are found by computing the determinant of certain submatrices.

A principal submatrix is a square submatrix obtained by removing certain rows and columns. The definition varies from author to author. According to some authors, a principal submatrix is a submatrix in which the set of row indices that remain is the same as the set of column indices that remain. Other authors define a principal submatrix to be one in which the first k rows and columns, for some number k, are the ones that remain; this type of submatrix has also been called a leading principal submatrix.

Linear Equations

Matrices can be used to compactly write and work with multiple linear equations, that is, systems of linear equations. For example, if A is an m-by-n matrix, x designates a column vector (that is, n×1-matrix) of n variables $x_1, x_2, \ldots, x_n$, and b is an m×1-column vector, then the matrix equation

$$Ax \quad b$$

is equivalent to the system of linear equations

$$A_{1,1}x_1 + A_{1,2}x_2 + \dots + A_{1,n}x_n = b_1$$

$$\dots$$

$$A_{m,1}x_1 + A_{m,2}x_2 + \dots + A_{m,n}x_n = b_m .$$

Using matrices, this can be solved more compactly than would be possible by writing out all the equations separately. If $n = m$ and the equations are independent, this can be done by writing

$$x = A^{-1}b,$$

where A^{-1} is the inverse matrix of A. If A has no inverse, solutions if any can be found using its generalized inverse.

Linear Transformations

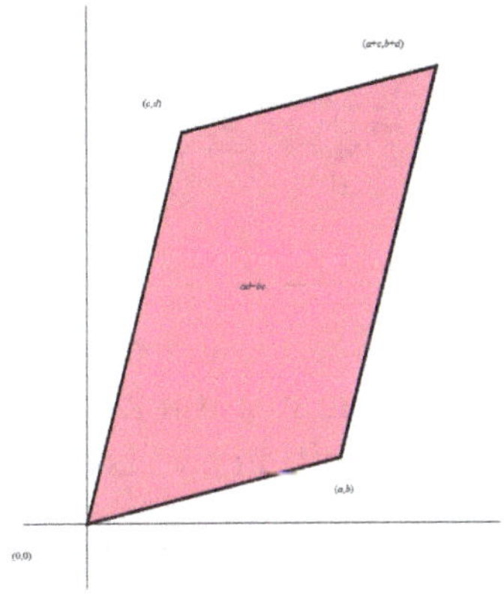

The vectors represented by a 2-by-2 matrix correspond to the sides of a unit square transformed into a parallelogram.

Matrices and matrix multiplication reveal their essential features when related to *linear transformations*, also known as *linear maps*. A real m-by-n matrix **A** gives rise to a linear transformation $R^n \to R^m$ mapping each vector x in R^n to the (matrix) product Ax, which is a vector in R^m. Conversely, each linear transformation f: $R^n \to R^m$ arises from a unique m-by-n matrix A: explicitly, the (i, j)-entry of A is the ith coordinate of $f(e_j)$, where $e_j = (0,\dots,0,1,0,\dots,0)$ is the unit vector with 1 in the jth position and 0 elsewhere. The matrix A is said to represent the linear map *f*, and **A** is called the *transformation matrix* of *f*.

For example, the 2×2 matrix

$$A = \begin{bmatrix} a & c \\ b & d \end{bmatrix}$$

can be viewed as the transform of the unit square into a parallelogram with vertices at (0, 0), (a, b), $(a + c, b + d)$, and (c, d). The parallelogram pictured at the right is obtained by multiplying **A** with each of the column vectors $\begin{bmatrix}0\\0\end{bmatrix}, \begin{bmatrix}1\\0\end{bmatrix}, \begin{bmatrix}1\\1\end{bmatrix}$ and $\begin{bmatrix}0\\1\end{bmatrix}$ in turn. These vectors define the vertices of the unit square.

The following table shows a number of 2-by-2 matrices with the associated linear maps of R^2. The blue original is mapped to the green grid and shapes. The origin (0,0) is marked with a black point.

Horizontal shear with m=1.25.	Reflection through the vertical axis	Squeeze mapping with r=3/2	Scaling by a factor of 3/2	Rotation by $\pi/6^R$ = 30°
$\begin{bmatrix}1 & 1.25\\0 & 1\end{bmatrix}$	$\begin{bmatrix}-1 & 0\\0 & 1\end{bmatrix}$	$\begin{bmatrix}3/2 & 0\\0 & 2/3\end{bmatrix}$	$\begin{bmatrix}3/2 & 0\\0 & 3/2\end{bmatrix}$	$\begin{bmatrix}\cos(\pi/6^R) & -\sin(\pi/6^R)\\\sin(\pi/6^R) & \cos(\pi/6^R)\end{bmatrix}$
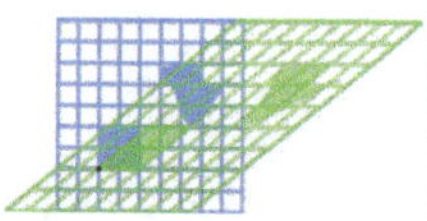	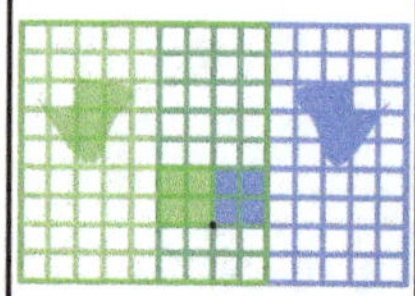	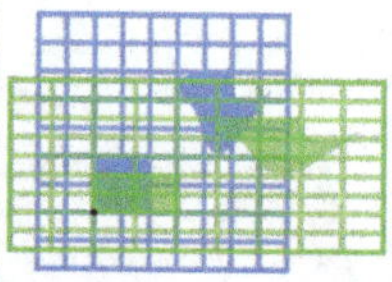	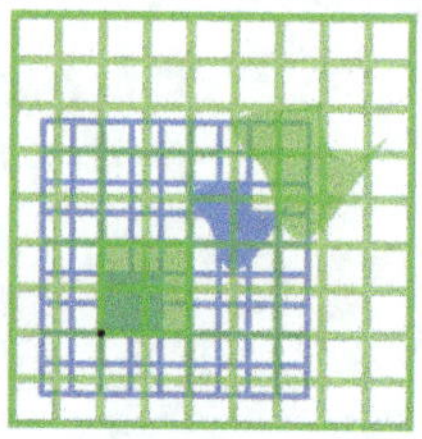	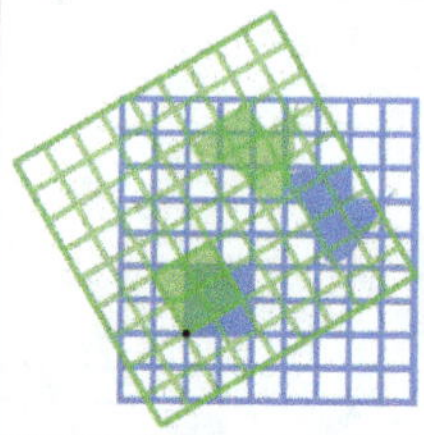

Under the 1-to-1 correspondence between matrices and linear maps, matrix multiplication corresponds to composition of maps: if a k-by-m matrix B represents another linear map $g : R^m \to R^k$, then the composition $g \circ f$ is represented by BA since

$$(g \circ f)(x) = g(f(x)) = g(Ax) = B(Ax) = (BA)x.$$

The last equality follows from the above-mentioned associativity of matrix multiplication.

The rank of a matrix **A** is the maximum number of linearly independent row vectors of the matrix, which is the same as the maximum number of linearly independent column vectors. Equivalently it is the dimension of the image of the linear map represented by **A**. The rank–nullity theorem states that the dimension of the kernel of a matrix plus the rank equals the number of columns of the matrix.

Square Matrix

A square matrix is a matrix with the same number of rows and columns. An n-by-n matrix is known as a square matrix of order n. Any two square matrices of the same order can be added and multiplied. The entries a_{ii} form the main diagonal of a square matrix. They lie on the imaginary line which runs from the top left corner to the bottom right corner of the matrix.

Main Types

Name	Example with $n = 3$
Diagonal matrix	$\begin{bmatrix} a_{11} & 0 & 0 \\ 0 & a_{22} & 0 \\ 0 & 0 & a_{33} \end{bmatrix}$
Lower triangular matrix	$\begin{bmatrix} a_{11} & 0 & 0 \\ a_{21} & a_{22} & 0 \\ a_{31} & a_{32} & a_{33} \end{bmatrix}$
Upper triangular matrix	$\begin{bmatrix} a_{11} & a_{12} & a_{13} \\ 0 & a_{22} & a_{23} \\ 0 & 0 & a_{33} \end{bmatrix}$

Diagonal and Triangular Matrix

If all entries of A below the main diagonal are zero, A is called an *upper triangular matrix*. Similarly if all entries of *A* above the main diagonal are zero, A is called a *lower triangular matrix*. If all entries outside the main diagonal are zero, A is called a diagonal matrix.

Identity Matrix

The *identity matrix* I_n of size *n* is the *n*-by-*n* matrix in which all the elements on the main diagonal are equal to 1 and all other elements are equal to 0, for example,

$$I_1 = [1], I_2 = \begin{bmatrix} 1 & 0 \\ 0 & 1 \end{bmatrix}, \cdots, I_n = \begin{bmatrix} 1 & 0 & \cdots & 0 \\ 0 & 1 & \cdots & 0 \\ \vdots & \vdots & \ddots & \vdots \\ 0 & 0 & \cdots & 1 \end{bmatrix}$$

It is a square matrix of order *n*, and also a special kind of diagonal matrix. It is called an identity matrix because multiplication with it leaves a matrix unchanged:

$AI_n = I_mA = A$ for any m-by-n matrix A.

A nonzero scalar multiple of an identity matrix is called a *scalar* matrix. If the matrix entries come from a field, the scalar matrices form a group, under matrix multiplication, that is isomorphic to the multiplicative group of nonzero elements of the field.

Symmetric or Skew-symmetric Matrix

A square matrix A that is equal to its transpose, that is, $A = A^T$, is a symmetric matrix. If instead, A is equal to the negative of its transpose, that is, $A = -A^T$, then A is a skew-symmetric matrix. In complex matrices, symmetry is often replaced by the concept of Hermitian matrices, which satisfy $A^* = A$, where the star or asterisk denotes the conjugate transpose of the matrix, that is, the transpose of the complex conjugate of A.

By the spectral theorem, real symmetric matrices and complex Hermitian matrices have an eigenbasis; that is, every vector is expressible as a linear combination of eigenvectors. In both cases, all eigenvalues are real. This theorem can be generalized to infinite-dimensional situations related to matrices with infinitely many rows and columns.

Invertible Matrix and its Inverse

A square matrix A is called *invertible* or *non-singular* if there exists a matrix B such that

$AB = BA = I_n$,

where I_n is the $n\times n$ identity matrix with 1s on the main diagonal and 0s elsewhere. If B exists, it is unique and is called the *inverse matrix* of A, denoted A^{-1}.

Definite Matrix

Positive definite matrix	**Indefinite matrix**
$\begin{bmatrix} 1/4 & 0 \\ 0 & 1 \end{bmatrix}$	$\begin{bmatrix} 1/4 & 0 \\ 0 & -1/4 \end{bmatrix}$
$Q(x,y) = 1/4\, x^2 + y^2$	$Q(x,y) = 1/4\, x^2 - 1/4\, y^2$
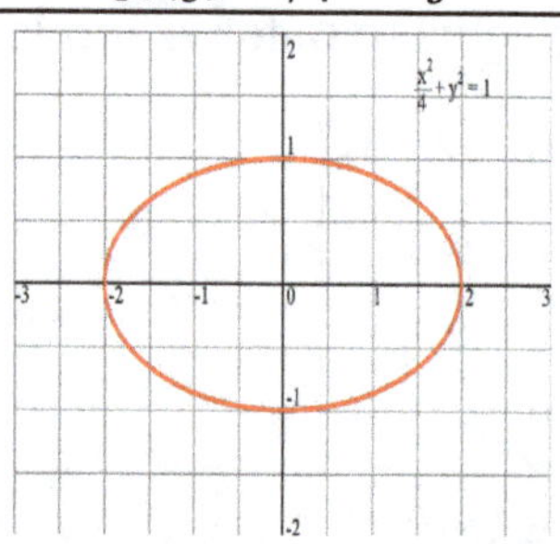 Points such that $Q(x,y)=1$ (Ellipse).	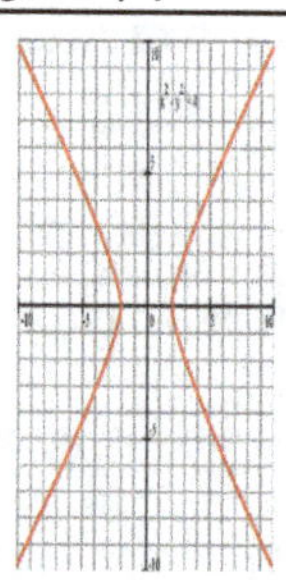 Points such that $Q(x,y)=1$ (Hyperbola).

A symmetric $n\times n$-matrix A is called positive-definite (respectively negative-definite; indefinite), if for all nonzero vectors $x \in R^n$ the associated quadratic form given by

$$Q(x) = x^T A x$$

takes only positive values (respectively only negative values; both some negative and some positive values). If the quadratic form takes only non-negative (respectively only non-positive) values, the symmetric matrix is called positive-semidefinite (respectively negative-semidefinite); hence the matrix is indefinite precisely when it is neither positive-semidefinite nor negative-semidefinite.

A symmetric matrix is positive-definite if and only if all its eigenvalues are positive, that is, the matrix is positive-semidefinite and it is invertible. The table shows two possibilities for 2-by-2 matrices.

Allowing as input two different vectors instead yields the bilinear form associated to **A**:

$$B_A(x, y) = x^T A y.$$

Orthogonal Matrix

An *orthogonal matrix* is a square matrix with real entries whose columns and rows are orthogonal unit vectors (that is, orthonormal vectors). Equivalently, a matrix A is orthogonal if its transpose is equal to its inverse:

$$A^T = A^{-1},$$

which entails

$$A^T A = AA^T = I_n,$$

where I is the identity matrix of size n.

An orthogonal matrix A is necessarily invertible (with inverse $A^{-1} = A^T$), unitary ($A^{-1} = A^*$), and normal ($A^*A = AA^*$). The determinant of any orthogonal matrix is either +1 or −1. A *special orthogonal matrix* is an orthogonal matrix with determinant +1. As a linear transformation, every orthogonal matrix with determinant +1 is a pure rotation, while every orthogonal matrix with determinant -1 is either a pure reflection, or a composition of reflection and rotation.

The complex analogue of an orthogonal matrix is a unitary matrix.

Main Operations

Trace

The trace, tr(**A**) of a square matrix **A** is the sum of its diagonal entries. While matrix multiplication is not commutative as mentioned above, the trace of the product of two matrices is independent of the order of the factors:

$$\mathrm{tr}(AB) = \mathrm{tr}(BA).$$

This is immediate from the definition of matrix multiplication:

$$\text{tr}(AB) = \sum_{i=1}^{m}\sum_{j=1}^{n} A_{ij}B_{ji} = \text{tr}(BA).$$

Also, the trace of a matrix is equal to that of its transpose, that is,

$$\text{tr}(A) = \text{tr}(A^T).$$

Determinant

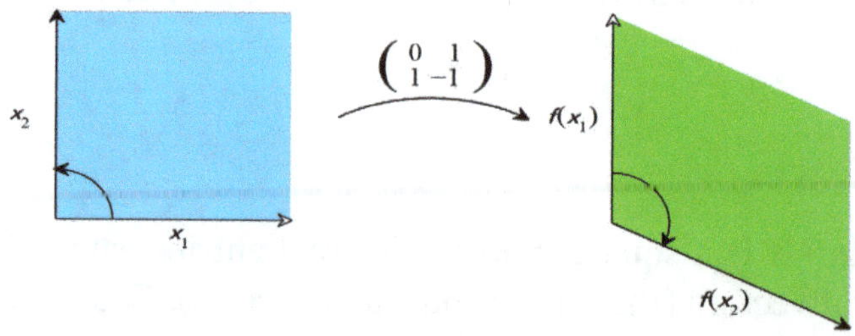

A linear transformation on R^2 given by the indicated matrix. The determinant of this matrix is −1, as the area of the green parallelogram at the right is 1, but the map reverses the orientation, since it turns the counterclockwise orientation of the vectors to a clockwise one.

The determinant det(A) or |A| of a square matrix A is a number encoding certain properties of the matrix. A matrix is invertible if and only if its determinant is nonzero. Its absolute value equals the area (in R^2) or volume (in R^3) of the image of the unit square (or cube), while its sign corresponds to the orientation of the corresponding linear map: the determinant is positive if and only if the orientation is preserved.

The determinant of 2-by-2 matrices is given by

$$\det\begin{bmatrix} a & b \\ c & d \end{bmatrix} = ad - bc.$$

The determinant of 3-by-3 matrices involves 6 terms (rule of Sarrus). The more lengthy Leibniz formula generalises these two formulae to all dimensions.

The determinant of a product of square matrices equals the product of their determinants:

$$\det(AB) = \det(A) \cdot \det(B).$$

Adding a multiple of any row to another row, or a multiple of any column to another column, does not change the determinant. Interchanging two rows or two columns affects the determinant by multiplying it by −1. Using these operations, any matrix can be transformed to a lower (or upper) triangular matrix, and for such matrices the determinant equals the product of the entries on the main diagonal; this provides a method to calculate the determinant of any matrix. Finally, the Laplace expansion expresses the determinant in terms of minors, that is, determinants of smaller matrices. This expansion can be used for a recursive definition of determinants (taking as starting case the determinant of a 1-by-1 matrix, which is its unique entry, or even the determinant of a 0-by-0 matrix, which is 1), that can be seen to be equivalent to the Leibniz formula. Determinants can be used to solve linear systems using Cramer's rule, where the division of the determinants of two related square matrices equates to the value of each of the system's variables.

Eigenvalues and Eigenvectors

A number λ and a non-zero vector **v** satisfying

$$Av = \lambda v$$

are called an *eigenvalue* and an *eigenvector* of A, respectively. The number λ is an eigenvalue of an n×n-matrix A if and only if $A-\lambda I_n$ is not invertible, which is equivalent to

$$\det(A - \lambda I) = 0.$$

The polynomial p_A in an indeterminate *X* given by evaluation the determinant $\det(XI_n-A)$ is called the characteristic polynomial of A. It is a monic polynomial of degree n. Therefore the polynomial equation $p_A(\lambda) = 0$ has at most n different solutions, that is, eigenvalues of the matrix. They may be complex even if the entries of A are real. According to the Cayley–Hamilton theorem, $p_A(A) = \mathbf{0}$, that is, the result of substituting the matrix itself into its own characteristic polynomial yields the zero matrix.

Computational Aspects

Matrix calculations can be often performed with different techniques. Many problems can be solved by both direct algorithms or iterative approaches. For example, the eigenvectors of a square matrix can be obtained by finding a sequence of vectors $\mathbf{x}_n$ converging to an eigenvector when *n* tends to infinity.

To be able to choose the more appropriate algorithm for each specific problem, it is important to determine both the effectiveness and precision of all the available algorithms. The domain studying these matters is called numerical linear algebra. As with other numerical situations, two main aspects are the complexity of algorithms and their numerical stability.

Determining the complexity of an algorithm means finding upper bounds or estimates of how many elementary operations such as additions and multiplications of scalars are necessary to perform some algorithm, for example, multiplication of matrices. For example, calculating the matrix product of two *n*-by-*n* matrix using the definition given above needs n^3 multiplications, since for any of the n^2 entries of the product, *n* multiplications are necessary. The Strassen algorithm outperforms this "naive" algorithm; it needs only $n^{2.807}$ multiplications. A refined approach also incorporates specific features of the computing devices.

In many practical situations additional information about the matrices involved is known. An important case are sparse matrices, that is, matrices most of whose entries are zero. There are specifically adapted algorithms for, say, solving linear systems Ax = b for sparse matrices A, such as the conjugate gradient method.

An algorithm is, roughly speaking, numerically stable, if little deviations in the input values do not lead to big deviations in the result. For example, calculating the inverse of a matrix via Laplace's formula (Adj (A) denotes the adjugate matrix of A)

$$A^{-1} = \mathrm{Adj}(A) / \det(A)$$

may lead to significant rounding errors if the determinant of the matrix is very small. The norm of a matrix can be used to capture the conditioning of linear algebraic problems, such as computing a matrix's inverse.

Although most computer languages are not designed with commands or libraries for matrices, as early as the 1970s, some engineering desktop computers such as the HP 9830 had ROM cartridges to add BASIC commands for matrices. Some computer languages such as APL were designed to manipulate matrices, and various mathematical programs can be used to aid computing with matrices.

Decomposition

There are several methods to render matrices into a more easily accessible form. They are generally referred to as *matrix decomposition* or *matrix factorization* techniques. The interest of all these techniques is that they preserve certain properties of the matrices in question, such as determinant, rank or inverse, so that these quantities can be calculated after applying the transformation, or that certain matrix operations are algorithmically easier to carry out for some types of matrices.

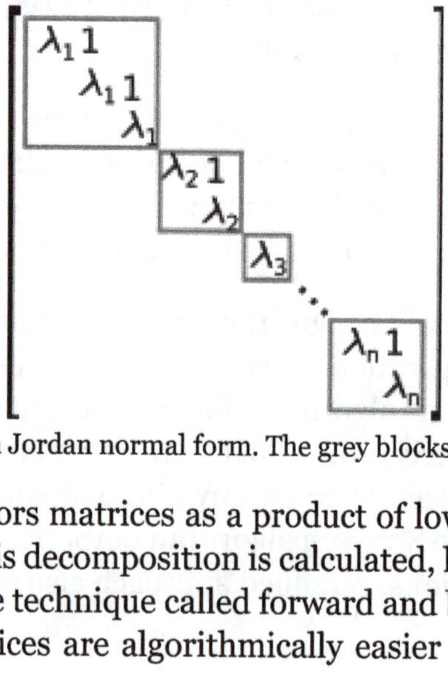

An example of a matrix in Jordan normal form. The grey blocks are called Jordan blocks.

The LU decomposition factors matrices as a product of lower (**L**) and an upper triangular matrices (**U**). Once this decomposition is calculated, linear systems can be solved more efficiently, by a simple technique called forward and back substitution. Likewise, inverses of triangular matrices are algorithmically easier to calculate. The *Gaussian* *elimination* is a similar algorithm; it transforms any matrix to row echelon form. Both methods proceed by multiplying the matrix by suitable elementary matrices, which correspond to permuting rows or columns and adding multiples of one row to another row. Singular value decomposition expresses any matrix A as a product UDV^*, where U and V are unitary matrices and D is a diagonal matrix.

The eigendecomposition or *diagonalization* expresses A as a product VDV^{-1}, where D is a diagonal matrix and V is a suitable invertible matrix. If A can be written in this form, it is called diagonalizable. More generally, and applicable to all matrices, the Jordan decomposition transforms a matrix into Jordan normal form, that is to say matrices whose only nonzero entries are the eigenvalues λ_1 to λ_n of A, placed on the main diagonal and possibly entries equal to one directly above the main diagonal. Given the eigendecomposition, the nth power of A (that is, n-fold iterated matrix multiplication) can be calculated via

$$A^n = (VDV^{-1})^n = VDV^{-1}VDV^{-1}\ldots VDV^{-1} = VD^nV^{-1}$$

and the power of a diagonal matrix can be calculated by taking the corresponding powers of the diagonal entries, which is much easier than doing the exponentiation for **A** instead. This can be used to compute the matrix exponential e^{A}, a need frequently arising in solving linear differential equations, matrix logarithms and square roots of matrices. To avoid numerically ill-conditioned situations, further algorithms such as the Schur decomposition can be employed.

Abstract Algebraic Aspects and Generalizations

Matrices can be generalized in different ways. Abstract algebra uses matrices with entries in more general fields or even rings, while linear algebra codifies properties of matrices in the notion of linear maps. It is possible to consider matrices with infinitely many columns and rows. Another extension are tensors, which can be seen as higher-dimensional arrays of numbers, as opposed to vectors, which can often be realised as sequences of numbers, while matrices are rectangular or two-dimensional arrays of numbers. Matrices, subject to certain requirements tend to form groups known as matrix groups. Similarly under certain conditions matrices form rings known as matrix rings. Though the product of matrices is not in general commutative yet certain matrices form fields known as matrix fields.

Matrices with More General Entries

This article focuses on matrices whose entries are real or complex numbers. However, matrices can be considered with much more general types of entries than real or complex numbers. As a first step of generalization, any field, that is, a set where addition, subtraction, multiplication and division operations are defined and well-behaved, may be used instead of R or C, for example rational numbers or finite fields. For example, coding theory makes use of matrices over finite fields. Wherever eigenvalues are considered, as these are roots of a polynomial they may exist only in a larger field than that of the entries of the matrix; for instance they may be complex in case of a matrix with real entries. The possibility to reinterpret the entries of a matrix as elements of a larger field (for example, to view a real matrix as a complex matrix whose entries happen to be all real) then allows considering each square matrix to possess a full set of eigenvalues. Alternatively one can consider only matrices with entries in an algebraically closed field, such as C, from the outset.

More generally, abstract algebra makes great use of matrices with entries in a ring *R*. Rings are a more general notion than fields in that a division operation need not exist. The very same addition and multiplication operations of matrices extend to this setting, too. The set $M(n, R)$ of all square *n*-by-*n* matrices over *R* is a ring called matrix ring, isomorphic to the endomorphism ring of the left *R*-module R^n. If the ring *R* is commutative, that is, its multiplication is commutative, then $M(n, R)$ is a unitary noncommutative (unless $n = 1$) associative algebra over *R*. The determinant of square matrices over a commutative ring *R* can still be defined using the Leibniz formula; such a matrix is invertible if and only if its determinant is invertible in *R*, generalising the situation over a field *F*, where every nonzero element is invertible. Matrices over superrings are called supermatrices.

Matrices do not always have all their entries in the same ring – or even in any ring at all. One special but common case is block matrices, which may be considered as matrices whose entries themselves are matrices. The entries need not be quadratic matrices, and thus need not be members of any ordinary ring; but their sizes must fulfil certain compatibility conditions.

Relationship to Linear Maps

Linear maps $R^n \to R^m$ are equivalent to m-by-n matrices, as described above. More generally, any linear map f: V → W between finite-dimensional vector spaces can be described by a matrix $A = (a_{ij})$, after choosing bases $v_1, \ldots, v_n$ of V, and $w_1, \ldots, w_m$ of *W* (so *n* is the dimension of *V* and *m* is the dimension of *W*), which is such that

$$f(v_j) = \sum_{i=1}^{m} a_{i,j} w_i \qquad \text{for } j = 1, \ldots, n.$$

In other words, column *j* of *A* expresses the image of $\mathbf{v}_j$ in terms of the basis vectors $\mathbf{w}_i$ of *W*; thus this relation uniquely determines the entries of the matrix **A**. Note that the matrix depends on the choice of the bases: different choices of bases give rise to different, but equivalent matrices. Many of the above concrete notions can be reinterpreted in this light, for example, the transpose matrix A^T describes the transpose of the linear map given by A, with respect to the dual bases.

These properties can be restated in a more natural way: the category of all matrices with entries in a field k with multiplication as composition is equivalent to the category of finite dimensional vector spaces and linear maps over this field.

More generally, the set of $m \times n$ matrices can be used to represent the R-linear maps between the free modules R^m and R^n for an arbitrary ring R with unity. When $n = m$ composition of these maps is possible, and this gives rise to the matrix ring of $n \times n$ matrices representing the endomorphism ring of R^n.

Matrix Groups

A group is a mathematical structure consisting of a set of objects together with a binary operation, that is, an operation combining any two objects to a third, subject to certain requirements. A group in which the objects are matrices and the group operation is matrix multiplication is called a *matrix group*. Since in a group every element has to be invertible, the most general matrix groups are the groups of all invertible matrices of a given size, called the general linear groups.

Any property of matrices that is preserved under matrix products and inverses can be used to define further matrix groups. For example, matrices with a given size and with a determinant of 1 form a subgroup of (that is, a smaller group contained in) their general linear group, called a special linear group. Orthogonal matrices, determined by the condition

$$M^T M = I,$$

form the orthogonal group. Every orthogonal matrix has determinant 1 or −1. Orthogonal matrices with determinant 1 form a subgroup called *special orthogonal group*.

Every finite group is isomorphic to a matrix group, as one can see by considering the regular representation of the symmetric group. General groups can be studied using matrix groups, which are comparatively well-understood, by means of representation theory.

Infinite Matrices

It is also possible to consider matrices with infinitely many rows and/or columns even if, being infinite objects, one cannot write down such matrices explicitly. All that matters is that for every element in the set indexing rows, and every element in the set indexing columns, there is a well-defined entry (these index sets need not even be subsets of the natural numbers). The basic operations of addition, subtraction, scalar multiplication and transposition can still be defined without problem; however matrix multiplication may involve infinite summations to define the resulting entries, and these are not defined in general.

If R is any ring with unity, then the ring of endomorphisms of $M = \bigoplus_{i \in I} R$ as a right R module is isomorphic to the ring of column finite matrices $\mathbb{CFM}_I(R)$ whose entries are indexed by $I \times I$, and whose columns each contain only finitely many nonzero entries. The endomorphisms of *M* considered as a left *R* module result in an analogous object, the row finite matrices $\mathbb{RFM}_I(R)$ whose rows each only have finitely many nonzero entries.

If infinite matrices are used to describe linear maps, then only those matrices can be used all of whose columns have but a finite number of nonzero entries, for the following reason. For a matrix A to describe a linear map f: V→W, bases for both spaces must have been chosen; recall that by definition this means that every vector in the space can be written uniquely as a (finite) linear combination of basis vectors, so that written as a (column) vector v of coefficients, only finitely many entries v_i are nonzero. Now the columns of A describe the images by f of individual basis vectors of V in the basis of W, which is only meaningful if these columns have only finitely many nonzero entries. There is no restriction on the rows of A however: in the product A·v there are only finitely many nonzero coefficients of v involved, so every one of its entries, even if it is given as an infinite sum of products, involves only finitely many nonzero terms and is therefore well defined. Moreover, this amounts to forming a linear combination of the columns of A that effectively involves only finitely many of them, whence the result has only finitely many nonzero entries, because each of those columns do. One also sees that products of two matrices of the given type is well defined (provided as usual that the column-index and row-index sets match), is again of the same type, and corresponds to the composition of linear maps.

If *R* is a normed ring, then the condition of row or column finiteness can be relaxed. With the norm in place, absolutely convergent series can be used instead of finite sums. For example, the matrices whose column sums are absolutely convergent sequences form a ring. Analogously of course, the matrices whose row sums are absolutely convergent series also form a ring.

In that vein, infinite matrices can also be used to describe operators on Hilbert spaces, where convergence and continuity questions arise, which again results in certain constraints that have to be imposed. However, the explicit point of view of matrices tends to obfuscate the matter, and the abstract and more powerful tools of functional analysis can be used instead.

Empty Matrices

An *empty matrix* is a matrix in which the number of rows or columns (or both) is zero. Empty matrices help dealing with maps involving the zero vector space. For example, if *A* is a 3-by-0 matrix and *B* is a 0-by-3 matrix, then *AB* is the 3-by-3 zero matrix corresponding to the null map from a 3-dimensional space *V* to itself, while *BA* is a 0-by-0 matrix. There is no common notation for empty matrices, but most computer algebra systems allow creating and computing with them. The determinant of the 0-by-0 matrix is 1 as follows from regarding the empty product occurring in the Leibniz formula for the determinant as 1. This value is also consistent with the fact that the identity map from any finite dimensional space to itself has determinant 1, a fact that is often used as a part of the characterization of determinants.

Applications

There are numerous applications of matrices, both in mathematics and other sciences. Some of them merely take advantage of the compact representation of a set of numbers in a matrix. For example, in game theory and economics, the payoff matrix encodes the payoff for two players, depending on which out of a given (finite) set of alternatives the players choose. Text mining and automated thesaurus compilation makes use of document-term matrices such as tf-idf to track frequencies of certain words in several documents.

Complex numbers can be represented by particular real 2-by-2 matrices via

$$a + ib \leftrightarrow \begin{bmatrix} a & -b \\ b & a \end{bmatrix},$$

under which addition and multiplication of complex numbers and matrices correspond to each other. For example, 2-by-2 rotation matrices represent the multiplication with some complex number of absolute value 1, as above. A similar interpretation is possible for quaternions and Clifford algebras in general.

Early encryption techniques such as the Hill cipher also used matrices. However, due to the linear nature of matrices, these codes are comparatively easy to break. Computer graphics uses matrices both to represent objects and to calculate transformations of objects using affine rotation matrices to accomplish tasks such as projecting a three-dimensional object onto a two-dimensional screen, corresponding to a theoretical camera observation. Matrices over a polynomial ring are important in the study of control theory.

Chemistry makes use of matrices in various ways, particularly since the use of quantum theory to discuss molecular bonding and spectroscopy. Examples are the overlap matrix and the Fock matrix used in solving the Roothaan equations to obtain the molecular orbitals of the Hartree–Fock method.

Graph Theory

The adjacency matrix of a finite graph is a basic notion of graph theory. It records which vertices of the graph are connected by an edge. Matrices containing just two different values (1 and 0 meaning for example "yes" and "no", respectively) are called logical matrices. The distance (or cost) matrix contains information about distances of the edges.

These concepts can be applied to websites connected by hyperlinks or cities connected by roads etc., in which case (unless the connection network is extremely dense) the matrices tend to be sparse, that is, contain few nonzero entries. Therefore, specifically tailored matrix algorithms can be used in network theory.

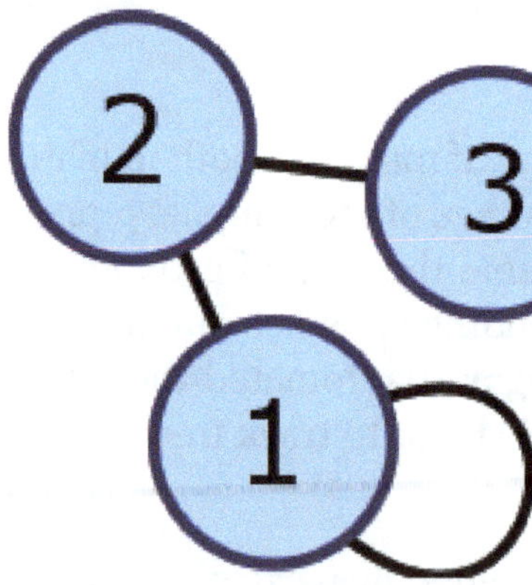

An undirected graph with adjacency matrix $\begin{bmatrix} 1 & 1 & 0 \\ 1 & 0 & 1 \\ 0 & 1 & 0 \end{bmatrix}$.

Analysis and Geometry

The Hessian matrix of a differentiable function f: $\mathbb{R}^n \rightarrow \mathbb{R}$ consists of the second derivatives of f with respect to the several coordinate directions, that is,

$$H(f) = \left[\frac{\partial^2 f}{\partial x_i \partial x_j} \right].$$

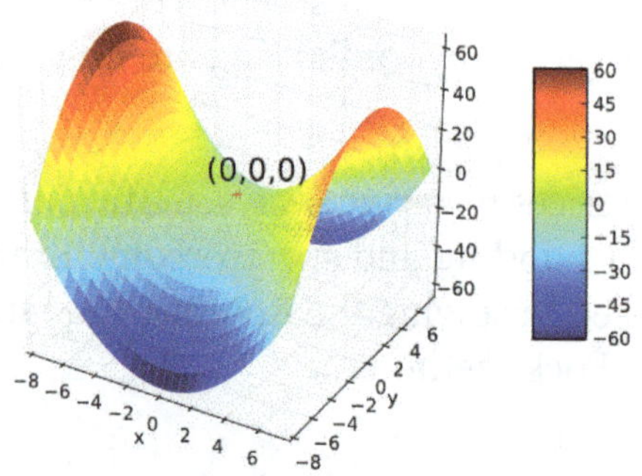

At the saddle point ($x = 0$, $y = 0$) (red) of the function $f(x,-y) = x^2 - y^2$, the Hessian matrix $\begin{bmatrix} 2 & 0 \\ 0 & -2 \end{bmatrix}$ is indefinite.

It encodes information about the local growth behaviour of the function: given a critical point $x = (x_1, \ldots, x_n)$, that is, a point where the first partial derivatives $\partial f / \partial x_i$ of

f vanish, the function has a local minimum if the Hessian matrix is positive definite. Quadratic programming can be used to find global minima or maxima of quadratic functions closely related to the ones attached to matrices.

Another matrix frequently used in geometrical situations is the Jacobi matrix of a differentiable map $f\colon R^n \to R^m$. If $f_1, \ldots, f_m$ denote the components of f, then the Jacobi matrix is defined as

$$J(f) = \left[\frac{\partial f_i}{\partial x_j}\right]_{1\le i\le m, 1\le j\le n}.$$

If $n > m$, and if the rank of the Jacobi matrix attains its maximal value m, f is locally invertible at that point, by the implicit function theorem.

Partial differential equations can be classified by considering the matrix of coefficients of the highest-order differential operators of the equation. For elliptic partial differential equations this matrix is positive definite, which has decisive influence on the set of possible solutions of the equation in question.

The finite element method is an important numerical method to solve partial differential equations, widely applied in simulating complex physical systems. It attempts to approximate the solution to some equation by piecewise linear functions, where the pieces are chosen with respect to a sufficiently fine grid, which in turn can be recast as a matrix equation.

Probability Theory and Statistics

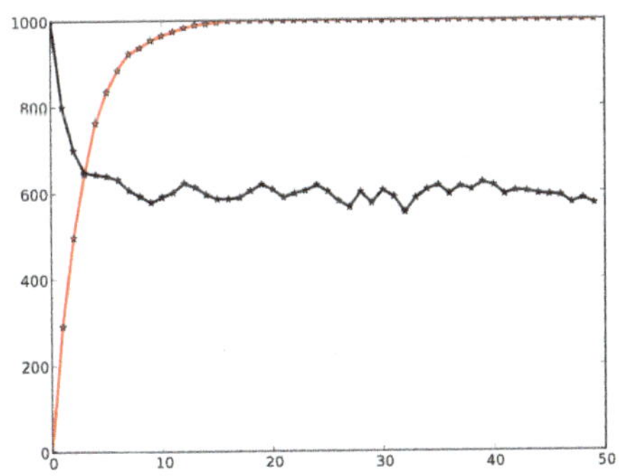

Two different Markov chains. The chart depicts the number of particles (of a total of 1000) in state "2". Both limiting values can be determined from the transition matrices, which are given by $\begin{bmatrix} .7 & 0 \\ .3 & 1 \end{bmatrix}$ (red) and $\begin{bmatrix} .7 & .2 \\ .3 & .8 \end{bmatrix}$ (black).

Stochastic matrices are square matrices whose rows are probability vectors, that is, whose entries are non-negative and sum up to one. Stochastic matrices are used to define Markov chains with finitely many states. A row of the stochastic matrix gives the probability distribution for the next position of some particle currently in the state that corresponds to the row. Properties of the Markov chain like absorbing states, that is, states that any particle attains eventually, can be read off the eigenvectors of the transition matrices.

Statistics also makes use of matrices in many different forms. Descriptive statistics is concerned with describing data sets, which can often be represented as data matrices, which may then be subjected to dimensionality reduction techniques. The covariance matrix encodes the mutual variance of several random variables. Another technique using matrices are linear least squares, a method that approximates a finite set of pairs $(x_1, y_1), (x_2, y_2), \ldots, (x_N, y_N)$, by a linear function

$$y_i \approx ax_i + b, i = 1, \ldots, N$$

which can be formulated in terms of matrices, related to the singular value decomposition of matrices.

Random matrices are matrices whose entries are random numbers, subject to suitable probability distributions, such as matrix normal distribution. Beyond probability theory, they are applied in domains ranging from number theory to physics.

Symmetries and Transformations in Physics

Linear transformations and the associated symmetries play a key role in modern physics. For example, elementary particles in quantum field theory are classified as representations of the Lorentz group of special relativity and, more specifically, by their behavior under the spin group. Concrete representations involving the Pauli matrices and more general gamma matrices are an integral part of the physical description of fermions, which behave as spinors. For the three lightest quarks, there is a group-theoretical representation involving the special unitary group SU(3); for their calculations, physicists use a convenient matrix representation known as the Gell-Mann matrices, which are also used for the SU(3) gauge group that forms the basis of the modern description of strong nuclear interactions, quantum chromodynamics. The Cabibbo–Kobayashi–Maskawa matrix, in turn, expresses the fact that the basic quark states that are important for weak interactions are not the same as, but linearly related to the basic quark states that define particles with specific and distinct masses.

Linear Combinations of Quantum States

The first model of quantum mechanics (Heisenberg, 1925) represented the theory's operators by infinite-dimensional matrices acting on quantum states. This is also referred to as matrix mechanics. One particular example is the density matrix that characterizes the "mixed" state of a quantum system as a linear combination of elementary, "pure" eigenstates.

Another matrix serves as a key tool for describing the scattering experiments that form the cornerstone of experimental particle physics: Collision reactions such as occur in particle accelerators, where non-interacting particles head towards each other and collide in a small interaction zone, with a new set of non-interacting particles as the result, can be described as the scalar product of outgoing particle states and a linear combination of ingoing particle states. The linear combination is given by a matrix known as the S-matrix, which encodes all information about the possible interactions between particles.

Normal Modes

A general application of matrices in physics is to the description of linearly coupled harmonic systems. The equations of motion of such systems can be described in matrix form, with a mass matrix multiplying a generalized velocity to give the kinetic term, and a force matrix multiplying a displacement vector to characterize the interactions. The best way to obtain solutions is to determine the system's eigenvectors, its normal modes, by diagonalizing the matrix equation. Techniques like this are crucial when it comes to the internal dynamics of molecules: the internal vibrations of systems consisting of mutually bound component atoms. They are also needed for describing mechanical vibrations, and oscillations in electrical circuits.

Geometrical Optics

Geometrical optics provides further matrix applications. In this approximative theory, the wave nature of light is neglected. The result is a model in which light rays are indeed geometrical rays. If the deflection of light rays by optical elements is small, the action of a lens or reflective element on a given light ray can be expressed as multiplication of a two-component vector with a two-by-two matrix called ray transfer matrix: the vector's components are the light ray's slope and its distance from the optical axis, while the matrix encodes the properties of the optical element. Actually, there are two kinds of matrices, viz. a *refraction matrix* describing the refraction at a lens surface, and a *translation matrix*, describing the translation of the plane of reference to the next refracting surface, where another refraction matrix applies. The optical system, consisting of a combination of lenses and/or reflective elements, is simply described by the matrix resulting from the product of the components' matrices.

Electronics

Traditional mesh analysis and nodal analysis in electronics lead to a system of linear equations that can be described with a matrix.

The behaviour of many electronic components can be described using matrices. Let A be a 2-dimensional vector with the component's input voltage v_1 and input current i_1 as its elements, and let B be a 2-dimensional vector with the component's output voltage

v_2 and output current i_2 as its elements. Then the behaviour of the electronic component can be described by $B = H \cdot A$, where H is a 2 x 2 matrix containing one impedance element (h_{12}), one admittance element (h_{21}) and two dimensionless elements (h_{11} and h_{22}). Calculating a circuit now reduces to multiplying matrices.

History

Matrices have a long history of application in solving linear equations but they were known as arrays until the 1800s. The Chinese text *The Nine Chapters on the Mathematical Art* written in 10th–2nd century BCE is the first example of the use of array methods to solve simultaneous equations, including the concept of determinants. In 1545 Italian mathematician Girolamo Cardano brought the method to Europe when he published *Ars Magna*. The Japanese mathematician Seki used the same array methods to solve simultaneous equations in 1683. The Dutch Mathematician Jan de Witt represented transformations using arrays in his 1659 book *Elements of Curves* (1659). Between 1700 and 1710 Gottfried Wilhelm Leibniz publicized the use of arrays for recording information or solutions and experimented with over 50 different systems of arrays. Cramer presented his rule in 1750.

The term "matrix" (Latin for "womb", derived from *mater*—mother) was coined by James Joseph Sylvester in 1850, who understood a matrix as an object giving rise to a number of determinants today called minors, that is to say, determinants of smaller matrices that derive from the original one by removing columns and rows. In an 1851 paper, Sylvester explains:

> I have in previous papers defined a "Matrix" as a rectangular array of terms, out of which different systems of determinants may be engendered as from the womb of a common parent.

Arthur Cayley published a treatise on geometric transformations using matrices that were not rotated versions of the coefficients being investigated as had previously been done. Instead he defined operations such as addition, subtraction, multiplication, and division as transformations of those matrices and showed the associative and distributive properties held true. Cayley investigated and demonstrated the non-commutative property of matrix multiplication as well as the commutative property of matrix addition. Early matrix theory had limited the use of arrays almost exclusively to determinants and Arthur Cayley's abstract matrix operations were revolutionary. He was instrumental in proposing a matrix concept independent of equation systems. In 1858 Cayley published his *A memoir on the theory of matrices* in which he proposed and demonstrated the Cayley-Hamilton theorem.

An English mathematician named Cullis was the first to use modern bracket notation for matrices in 1913 and he simultaneously demonstrated the first significant use of the notation $A = [a_{i,j}]$ to represent a matrix where $a_{i,j}$ refers to the ith row and the jth column.

The study of determinants sprang from several sources. Number-theoretical problems led Gauss to relate coefficients of quadratic forms, that is, expressions such as $x^2 + xy - 2y^2$, and linear maps in three dimensions to matrices. Eisenstein further developed these notions, including the remark that, in modern parlance, matrix products are non-commutative. Cauchy was the first to prove general statements about determinants, using as definition of the determinant of a matrix $\mathbf{A} = [a_{i,j}]$ the following: replace the powers a_j^k by a_{jk} in the polynomial

$$a_1 a_2 \cdots a_n \prod_{i<j} (a_j - a_i),$$

where Π denotes the product of the indicated terms. He also showed, in 1829, that the eigenvalues of symmetric matrices are real. Jacobi studied "functional determinants"—later called Jacobi determinants by Sylvester—which can be used to describe geometric transformations at a local (or infinitesimal) level; Kronecker's *Vorlesungen über die Theorie der Determinanten* and Weierstrass' *Zur Determinantentheorie*, both published in 1903, first treated determinants axiomatically, as opposed to previous more concrete approaches such as the mentioned formula of Cauchy. At that point, determinants were firmly established.

Many theorems were first established for small matrices only, for example the Cayley–Hamilton theorem was proved for 2×2 matrices by Cayley in the aforementioned memoir, and by Hamilton for 4×4 matrices. Frobenius, working on bilinear forms, generalized the theorem to all dimensions (1898). Also at the end of the 19th century the Gauss–Jordan elimination (generalizing a special case now known as Gauss elimination) was established by Jordan. In the early 20th century, matrices attained a central role in linear algebra. partially due to their use in classification of the hypercomplex number systems of the previous century.

The inception of matrix mechanics by Heisenberg, Born and Jordan led to studying matrices with infinitely many rows and columns. Later, von Neumann carried out the mathematical formulation of quantum mechanics, by further developing functional analytic notions such as linear operators on Hilbert spaces, which, very roughly speaking, correspond to Euclidean space, but with an infinity of independent directions.

Other Historical Usages of the Word "Matrix" in Mathematics

The word has been used in unusual ways by at least two authors of historical importance.

Bertrand Russell and Alfred North Whitehead in their *Principia Mathematica* (1910–1913) use the word "matrix" in the context of their axiom of reducibility. They proposed this axiom as a means to reduce any function to one of lower type, successively, so that at the "bottom" (0 order) the function is identical to its extension:

> "Let us give the name of *matrix* to any function, of however many variables, which does not involve any apparent variables. Then any possible function other than a matrix is derived from a matrix by means of generalization, that is, by considering the proposition which asserts that the function in question is true with all possible values or with some value of one of the arguments, the other argument or arguments remaining undetermined".

For example, a function $\Phi(x, y)$ of two variables x and y can be reduced to a *collection* of functions of a single variable, for example, y, by "considering" the function for all possible values of "individuals" a_i substituted in place of variable x. And then the resulting collection of functions of the single variable y, that is, $\forall a_i$: $\Phi(a_i, y)$, can be reduced to a "matrix" of values by "considering" the function for all possible values of "individuals" b_i substituted in place of variable y:

$$\forall b_j \forall a_i : \Phi(a_i, b_j).$$

Alfred Tarski in his 1946 *Introduction to Logic* used the word "matrix" synonymously with the notion of truth table as used in mathematical logic.

2 × 2 real Matrices

In mathematics, the associative algebra of 2×2 real matrices is denoted by M(2, R). Two matrices p and q in M(2, R) have a sum $p + q$ given by matrix addition. The product matrix pq is formed from the dot product of the rows and columns of its factors through matrix multiplication. For

$$q = \begin{pmatrix} a & b \\ c & d \end{pmatrix},$$

let

$$q^* = \begin{pmatrix} d & -b \\ -c & a \end{pmatrix}.$$

Then $qq^* = q^*q = (ad - bc)\,I$, where I is the 2×2 identity matrix. The real number $ad - bc$ is called the determinant of q. When $ad - bc \neq 0$, q is an invertible matrix, and then

$$q^{-1} = q^*/(ad - bc).$$

The collection of all such invertible matrices constitutes the general linear group GL(2, R). In terms of abstract algebra, M(2, R) with the associated addition and multiplication operations forms a ring, and GL(2, R) is its group of units. M(2, R) is also a four-dimensional vector space, so it is considered an associative algebra. It is ring-isomorphic to the coquaternions, but has a different profile.

The 2×2 real matrices are in one-one correspondence with the linear mappings of the two-dimensional Cartesian coordinate system into itself by the rule

$$\begin{pmatrix} x \\ y \end{pmatrix} \mapsto \begin{pmatrix} a & b \\ c & d \end{pmatrix}\begin{pmatrix} x \\ y \end{pmatrix} = \begin{pmatrix} ax+by \\ cx+dy \end{pmatrix}.$$

Profile

Within M(2, R), the multiples by real numbers of the identity matrix I may be considered a real line. This real line is the place where all commutative subrings come together:

Let $P_m = \{xI + ym : x, y \in R\}$ where $m^2 \in \{-I, 0, I\}$. Then P_m is a commutative subring and $M(2, R) = \cup P_m$ where the union is over all m such that $m^2 \in \{-I, 0, I\}$.

To identify such m, first square the generic matrix:

$$\begin{pmatrix} aa+bc & ab+bd \\ ac+cd & bc+dd \end{pmatrix}.$$

When $a + d = 0$ this square is a diagonal matrix. Thus one assumes $d = -a$ when looking for m to form commutative subrings. When $mm = -I$, then $bc = -1 - aa$, an equation describing a hyperbolic paraboloid in the space of parameters (a, b, c). Such an m serves as an imaginary unit. In this case P_m is isomorphic to the field of (ordinary) complex numbers.

When $mm = +I$, m is an involutory matrix. Then $bc = +1 - aa$, also giving a hyperbolic paraboloid. If a matrix is an idempotent matrix, it must lie in such a P_m and in this case P_m is isomorphic to the ring of split-complex numbers.

The case of a nilpotent matrix, $mm = 0$, arises when only one of b or c is non-zero, and the commutative subring P_m is then a copy of the dual number plane.

When M(2, R) is reconfigured with a change of basis, this profile changes to the profile of split-quaternions where the sets of square roots of I and $-I$ take a symmetrical shape as hyperboloids.

Equi-areal Mapping

First transform one differential vector into another:

$$\begin{pmatrix} du \\ dv \end{pmatrix} = \begin{pmatrix} p & r \\ q & s \end{pmatrix}\begin{pmatrix} dx \\ dy \end{pmatrix} = \begin{pmatrix} pdx+rdy \\ qdx+sdy \end{pmatrix}.$$

Areas are measured with *density* $dx \wedge dy$, a differential 2-form which involves the use of exterior algebra. The transformed density is

$$du \wedge dv = 0 + ps\ dx \wedge dy + qr\ dy \wedge dx + 0$$
$$= (ps - qr)\ dx \wedge dy = (\det g)\ dx \wedge dy.$$

Thus the equi-areal mappings are identified with SL(2, R) = $\{g \in M(2, R) : \det(g) = 1\}$, the special linear group. Given the profile above, every such g lies in a commutative subring P_m representing a type of complex plane according to the square of m. Since $g\,g^* = I$, one of the following three alternatives occurs:

- $mm = -I$ and g is on a circle of Euclidean rotations; or
- $mm = I$ and g is on an hyperbola of squeeze mappings; or
- $mm = 0$ and g is on a line of shear mappings.

Writing about planar affine mapping, Rafael Artzy made a similar trichotomy of planar, linear mapping in his book *Linear Geometry* (1965).

Functions of 2 × 2 Real Matrices

The commutative subrings of M(2, R) determine the function theory; in particular the three types of subplanes have their own algebraic structures which set the value of algebraic expressions. Consideration of the square root function and the logarithm function serves to illustrate the constraints implied by the special properties of each type of subplane P_m described in the above profile. The concept of identity component of the group of units of P_m leads to the polar decomposition of elements of the group of units:

- If $mm = -I$, then $z = \rho \exp(\theta m)$.
- If $mm = 0$, then $z = \rho \exp(s\,m)$ or $z = -\rho \exp(s\,m)$.
- If $mm = I$, then $z = \rho \exp(a\,m)$ or $z = -\rho \exp(a\,m)$ or $z = m\,\rho \exp(a\,m)$ or $z = -m\,\rho \exp(a\,m)$.

In the first case $\exp(\theta\,m) = \cos(\theta) + m \sin(\theta)$. In the case of the dual numbers $\exp(s\,m) = 1 + s\,m$. Finally, in the case of split complex numbers there are four components in the group of units. The identity component is parameterized by ρ and $\exp(a\,m) = \cosh a + m \sinh a$.

Now $\sqrt{\rho \exp(am)} = \sqrt{\rho}\, \exp(am/2)$ regardless of the subplane P_m, but the argument of the function must be taken from the *identity component of its group of units*. Half the plane is lost in the case of the dual number structure; three-quarters of the plane must be excluded in the case of the split-complex number structure.

Similarly, if ρ exp(a m) is an element of the identity component of the group of units of a plane associated with 2×2 matrix m, then the logarithm function results in a value log ρ + a m. The domain of the logarithm function suffers the same constraints as does the square root function described above: half or three-quarters of P_m must be excluded in the cases $mm = 0$ or $mm = I$.

2 × 2 Real Matrices as Complex Numbers

Every 2×2 real matrix can be interpreted as one of three types of (generalized) complex numbers: standard complex numbers, dual numbers, and split-complex numbers. Above, the algebra of 2×2 matrices is profiled as a union of complex planes, all sharing the same real axis. These planes are presented as commutative subrings P_m. We can determine to which complex plane a given 2×2 matrix belongs as follows and classify which kind of complex number that plane represents.

Consider the 2×2 matrix

$$z = \begin{pmatrix} a & b \\ c & d \end{pmatrix}.$$

We seek the complex plane P_m containing z.

As noted above, the square of the matrix z is diagonal when $a + d = 0$. The matrix z must be expressed as the sum of a multiple of the identity matrix I and a matrix in the hyperplane $a + d = 0$. Projecting z alternately onto these subspaces of R^4 yields

$$z = xI + n, \quad x = \frac{a+d}{2}, \quad n = z - xI.$$

Furthermore,

$$n^2 = pI \text{ where } p = \frac{(a-d)^2}{4} + bc..$$

Now z is one of three types of complex number:

- If $p < 0$, then it is an ordinary complex number:

  Let $q = 1/\sqrt{-p}, \quad m = qn$. Then $m^2 = -I, \quad z = xI + m\sqrt{-p}$.

- If $p = 0$, then it is the dual number:

  $z = xI + n$.

- If $p > 0$, then z is a split-complex number:

  Let $q = 1/\sqrt{p}, \quad m = qn$. Then $m^2 = +I, \quad z = xI + m\sqrt{p}$.

Similarly, a 2×2 matrix can also be expressed in polar coordinates with the caveat that there are two connected components of the group of units in the dual number plane, and four components in the split-complex number plane.

Classical and Quantum Computation of Dual Basis

It is hard to imagine that by passing to a different basis, a quantum printer suddenly becomes a quantum scanner.

1. Recall that if X is a vector space of dim X = n < ∞ and Y is a vector space dim Y = m < ∞ then for a linear transformation $X \xrightarrow{A} Y$ the matrix of A with respect to chosen coordinatizations is given by

$$(X,(e,\epsilon)) \xrightarrow[a=\left[a_i^j\right]_{m\times n}]{A} (Y,(d,\delta))$$

where ϵ is the dual basis to the basis e of X, δ is the dual basis to the basis d of Y and $a_i^j = \langle \delta^j | Ae_i \rangle$ is the entry at the intersection of the j-th row and the i-th column of the matrix we have, for a given x ∈ X,

$$x = \sum_{i=0}^{n-1} e_k x^k,\ x^k = \langle \epsilon^k | x \rangle$$

$$y = Ax = \sum_{j=0}^{m-1} d_j y^j = \sum_{j=0}^{m-1} d_j \sum_{i=0}^{n-1} \langle \delta^j | Ae_i \rangle \langle \epsilon^i | x \rangle$$

$$\text{where } y^j = \langle \delta^j | y \rangle = \sum_{i=0}^{n-1} \langle \delta^j | Ae_i \rangle \langle \epsilon^i | x \rangle$$

We may use the symbol $[A]_e^d = a = \left[a_i^j\right]_{m\times n}$ for this matrix.

2. Now suppose X has two coordinate systems (e,ϵ) and (e',ϵ'). Then we have $e_i' = \sum_{k=0}^{n-1} e_k b_i^k$ for uniquely given $b_i^k \in \mathbb{F}$ and also $e_i = \sum_{k=0}^{n-1} e_k' (b')_i^k$ for uniquely given $(b')_i^k \in \mathbb{F}$.

(i) Writing $Be_i = e_i'$ provides a linear transformation $X \xrightarrow{B} X$. Since a linear transformation is determined its values on the basis vectors e_i. Then we have $e_i' = Be_i = \sum_{k=0}^{n-1} e_k \langle \epsilon^k | Be_i \rangle$ (look at equation above) so that $b_i^k = \langle \epsilon^k | Be_i \rangle$ (because in $e_i' = \sum_{k=0}^{n-1} e_k b_i^k$ the scalars $b - i^k$ are uniquely given). Thus the matrix of

$$(X,(e,\epsilon)) \xrightarrow{B} (X,(e,\epsilon))$$

is $b=\left[b_i^k\right]_{n\times n}$. At the same time, $Be_i^{'}=B\left(\sum_{k=0}^{n-1}e_k b_i^k\right)=\sum_{k=0}^{n-1}(Be_k)b_i^k=\sum_{k=0}^{n-1}e_k^{'}\left\langle(\epsilon')^k\middle|Be_i^{'}\right\rangle$

so that $b_i^k=\left\langle(\epsilon')^k\middle|Be_i^{'}\right\rangle$ also $b=\left[b_i^k\right]_{n\times n}$ is also the matrix of

$$(X,(e',\epsilon'))\xrightarrow{B}(X,(e',\epsilon'))$$

And of course, since $Be_i^{'}=e_i^{'}$ we have $b_i^k=\left\langle(\epsilon)^k\middle|e_i^{'}\right\rangle$ which says (look at equation above) that $b=\left[b_i^k\right]_{n\times n}$ is also the matrix of

$$(X,(e',\epsilon'))\xrightarrow{Id}(X,(e,\epsilon))$$

(ii) There are thus two ways of looking at the relation $e_i^{'}=Be_i$.

Passive There are two coordinate systems (Call them two observers); the source basis e = $\{e_i\}$ is changed into the target basis $e'=\{e_i^{'}\}$ (the observer (e',ϵ') replaces the observer (e,ϵ)). The old coordinate system (= the observer) (e,ϵ) measures the coordinates of a vector x(a point in the space) as $\left\langle(\epsilon)^k\middle|x\right\rangle=x^k$ and records it as $x=\sum_{k=0}^{n}\left|e_k\right\rangle\left\langle\epsilon^k\middle|x\right\rangle$

while the new coordinate system (e',ϵ') records it as $x=\sum_{k=0}^{n}\left|e_k^{'}\right\rangle\left\langle(\epsilon')^k\middle|x\right\rangle$ measuring its coordinates as $\left\langle(\epsilon')^k\middle|x\right\rangle$. This can be seen as $(X,(e',\epsilon'))\xrightarrow{Id}(X,(e,\epsilon))$ so that the vector x has not moved but the source basis e has been changed by the linear transformation B represented by the matrix $\left[b_i^k\right]_{n\times n}$ into the target basis e' and thus the matrix $\left[b_i^k\right]_{n\times n}$ can be called as the change of basis matrix; thus $b=\left[b_i^k\right]_{n\times n}$ works as follows: Given a vector $x'=\sum_{i=0}^{n-1}e_i^{'}(x')^i$ in terms of the target basis e', the application of b rewrites x' in terms of the source basis e

$$\begin{aligned}
x&=\sum_{i=0}^{n-1}e_i^{'}(x')^{i}{}'\\
&=\sum_{i=0}^{n-1}\left|e_i^{'}\right\rangle\left\langle(\epsilon')^i\middle|x'\right\rangle\\
&=\sum_{i=0}^{n-1}\sum_{k=0}^{n-1}\left|e_k\right\rangle\left\langle\epsilon^k\middle|Ide_i^{'}\right\rangle\left\langle(\epsilon')^i\middle|x'\right\rangle\\
&=\sum_{k=0}^{n-1}\left|e_k\right\rangle[\sum_{i=0}^{n-1}b_i^k(x')^i]
\end{aligned}$$

Let us note carefully that the arrow in $(X,(e',\epsilon'))\xrightarrow{Id}(X,(e,\epsilon))$ points from the 'target basis' e' to the 'source basis' 'e'. The 'source-target' vocabulary is with reference to the linear transformation B (e_i gets transformed to $e_i' = Be_i$). Thus we have $B_e^e = B_{e'}^{e'} = P_{e'}^e = [Id]_{e'}^e$ in the notation of paragraph. The other interpretation is

Active: There is only one coordinate system (= observer), say (e,ϵ), for X. The $x = \sum_{i=0}^{n-1} e_i x^i$ moves to a different vector $u = Bx = \sum_{i=0}^{n-1}(Be_i)x^i = \sum_{i=0}^{n-1} e_i' x^i$ and the new vector u has components, with respect to (e,ϵ)

$$u^k = \langle \epsilon^k | u \rangle = \sum_{i=0}^{n-1} \langle \epsilon^k | Be_i \rangle x^i$$

So that $u = \sum_{k=0}^{n-1}\sum_{i=0}^{n-1} e_k b_i^k x^i$.

(while of course, the k-th component of x is $\langle \epsilon^k | x \rangle = x^k$).

3. Taking the passive interpretation, let us write the change of basis matrix b from $\{e_i\}$ to $\{e_i' = Be_i\}$ as $P_{e'}^e$ so that $P_{e'}^e(e_i') = \sum_{i=0}^{n-1} e_k b_i^k$ (that is, e_i' has been written in terms of the $\{e_k\}$) where $b_i^k = \langle \epsilon^k | e_i \rangle$.

Example 4.1 Consider $\mathbb{R}^3 = X$ the three-dimensional vector space over $\mathbb{R}$. Take $x = \begin{bmatrix} a \\ b \\ c \end{bmatrix}$ in X

then if $e = \left\{ e_0 = \begin{bmatrix} 1 \\ 2 \\ 0 \end{bmatrix}, e_1 = \begin{bmatrix} 1 \\ 3 \\ 2 \end{bmatrix}, e_2 = \begin{bmatrix} 0 \\ 1 \\ 3 \end{bmatrix} \right\}$ we have

$x = \begin{bmatrix} a \\ b \\ c \end{bmatrix} = e_0(7a-3b+c) + e_1(-6a+3b-c) + e_2(4a-2b+c)$ saying that each x is expressible

uniquely as $\sum_{i=0}^{2} e_i'(x')^i$ as well and e' is also a basis. Then because

$e_0' = 2e_0 - e_1 + e - 2, e_1' = -e_0 + e_1 + 0e_2, e_2' = e_0 + 0e_1 + 2e_2$ the change of basis matrix

which changes e into e' is $P_{e'}^e = b = \begin{bmatrix} 2 & -1 & 1 \\ -1 & 1 & 0 \\ 1 & 0 & 2 \end{bmatrix}$ (Obtained by writing components of

e_0', e_1', e_2' with respect to e as columns).

$$P_{e'}^{e}\left(\sum_{i=0}^{2} e_i^{'}(x')^i\right)=\begin{bmatrix}2 & -1 & 1\\ -1 & 1 & 0\\ 1 & 0 & 2\end{bmatrix}\begin{bmatrix}-2a+2b-c\\ -8a+5b-2c\\ 3a-2b+c\end{bmatrix}=\begin{bmatrix}7a-3b+c\\ -6a+3b-c\\ 4a-2b+c\end{bmatrix}=\sum_{i=0}^{2} e_i x^i \text{ which confirms}$$

$(X,(e',\epsilon'))\xrightarrow[P_{e'}^{e}=b=\left[b_i^j\right]]{Id}(X,(e,\epsilon))$ in this case.

4. Consider now the matrix which changes e' into e this will be the matrix b' given by

$$(X,(e,\epsilon))\xrightarrow[P_{e'}^{e}=b'=\left[(b')_i^j\right]_{n\times n}]{Id}(X,(e',\epsilon'))$$

where $(b')_i^j=\left\langle(\epsilon')^j\middle|e_i\right\rangle$ where $e_i=\sum_{k=0}^{n-1} e_k^{'}(b')_i^k$ and we may write $B'e_i^{'}=e_i$ which provides a linear transformation $X\xrightarrow{B'}X$ ($\because e'=\{e_i^{'}\}$ is a basis and a linear transformation is determined by its values on the elements of a basis). Then $BB'e_i^{'}=Be_i=e_i^{'}$ and $B'Be_i=Id=B'B$. We of course write B^{-1} for B'.

Then $(b')_i^j=\left\langle(\epsilon')^j\middle|e_i\right\rangle=\left\langle(\epsilon')^j\middle|B'e_i^{'}\right\rangle$ presents b' as the matrix for B' as

$$(X,(e',\epsilon'))\xrightarrow[b'=\left[(b')_i^j\right]_{n\times n}]{B'}(X,(e',\epsilon'))$$

while $B'e_i=B'\left(\sum_{k=0}^{n-1} e_k^{'}(b')_i^k\right)=\sum_{k=0}^{n-1}(B'e_k^{'})(b')_i^k=\sum_{k=0}^{n-1} e_k(b')_i^k=\sum_{k=0}^{n-1} e_k\left\langle\epsilon^k\middle|B'e_i\right\rangle$

($\therefore$ $\sum_{k=0}^{n-1} e_k\left\langle\epsilon^k\middle|B'e_i\right\rangle$ being the unique representation in terms of the basis e for $B'e_i$ like any vector $x=\sum_{k=0}^{n-1} e_k\left\langle\epsilon^k\middle|x\right\rangle$)

we get $(b')_i^k=\left\langle\epsilon^k\middle|B'e_i\right\rangle$ which means b' is a matrix of

$$(X,(e,\epsilon))\xrightarrow[b'=\left[(b')_i^k\right]]{B'}(X,(e,\epsilon))$$

Example (continued) With the illustration above, we have

$$P_{e}^{e'}=\left(P_{e'}^{e}\right)^{-1}=b^{-1}=b'=\left[\left(b^{-1}\right)_i^j\right]=\begin{bmatrix}2 & 2 & -1\\ 2 & 3 & -1\\ -1 & -1 & 1\end{bmatrix}$$

which can be verified directly as above and we get

$$\begin{bmatrix} 2 & 2 & -1 \\ 2 & 3 & -1 \\ -1 & -1 & 1 \end{bmatrix}\begin{bmatrix} 7a+3b+c \\ -6a+3b-c \\ 4a-2b+c \end{bmatrix}=\begin{bmatrix} -2a+2b-c \\ -8a+5b-2c \\ 3a-2b+c \end{bmatrix} \text{ which confirms}$$

$$\left(X,(e',\epsilon')\right)\xrightarrow[P_e^{e'}=b^{-1}=b'=\left[\left(b^{-1}\right)_i^j\right]]{\text{Id}}\left(X,(e,\epsilon)\right) \quad \text{in this case.}$$

5. The preceding discussion shows that a change of basis matrix is invertible. Now take an invertible matrix $a=\left[a_i^j\right]$, then $e_i = a^{-1}ae_i$ shows that each

$$\begin{aligned} x &= \sum_{k=0}^{n-1} e_i x^i \\ &= \sum_{k=0}^{n-1} a^{-1}\left(ae_i\right)x^i \\ &= \sum_{i=0}^{n-1}\left(\sum_{k=0}^{n-1}\left(a^{-1}\right)_i^k\left(ae_k\right)\right)x^i \\ &= \sum_{i=0}^{n-1}\left(ae_k\right)\left(\sum_{k=0}^{n-1}\left(a^{-1}\right)_i^k x^i\right) \\ &= \sum_{i=0}^{n-1}\left(ae_k\right)\lambda^k \quad \text{(say)} \end{aligned}$$

showing that each x is expressible as uniquely as a linear combination of the vectors $\{ae_k\}$ constitute a basis and a is the change of basis matrix from the source basis $\{e_i\}$ to the target basis $\left\{e_i' = ae_i\right\}$. (To understand why we wrote $a^{-1}(ae_i)$ as $\sum_{i=0}^{n-1}\left(a^{-1}\right)_i^k\left(ae_k\right)$, note that when B' is a linear transformation, the formula is $B'\left(e_i'\right)=e_i=\sum_{i=0}^{n-1}e_k'\left(b'\right)_i^k$ here the application of a^{-1} produces a linear transformation B' and we have $ae_i = e_i'$ so that the matrix of B' here is $b'=a^{-1}$ itself and we have $e_i = a^{-1}\left(ae_i\right)=\sum_{i=0}^{n-1}\left(e_k'\right)\left(b'\right)_i^k=\sum_{i=0}^{n-1}\left(ae_k\right)\left(a^{-1}\right)_i^k$).

Thus a change of basis matrix is the same as an invertible matrix.

6. If $P_{e'}^{e}=b=\left[b_i^j\right]$ is the change of basis matrix changing e to e' the formula $e_i'=\sum_{j=0}^{n-1}\left(e_j\right)b_i^j$ says that we expand the vector e_i', that is, the i–th vector of the target basis e', in terms of the source basis e = $\{e_j\}$; the components in this expansion (namely, the scalars b_i^j) then supply the i-th column of the matrix $P_{e'}^{e}=b$ which is $b_i=\begin{bmatrix} b_i^0 \\ \vdots \\ b_i^{n-1} \end{bmatrix}$.

Example 0.1 Consider $\mathbb{K}=\mathbb{R},\ \mathbb{X}=\mathbb{R}^2$; then $e=\left\{e_0=\begin{bmatrix}1\\-2\end{bmatrix}, e_1=\begin{bmatrix}3\\-4\end{bmatrix}\right\}$ is a basis

$\left(\because\begin{bmatrix}a\\b\end{bmatrix}=e_0\left(-2a-\frac{3}{2}b\right)+e_1\left(a+\frac{1}{2}b\right)\text{uniquely expressed}\right)$ and $e'=\left\{e_0=\begin{bmatrix}1\\3\end{bmatrix}, e_1=\begin{bmatrix}3\\8\end{bmatrix}\right\}$

is also a basis $\left(\because\begin{bmatrix}a\\b\end{bmatrix}=e_0'(-8a-3b)+e_1'(3a-b)\text{uniquely expressed}\right)$.

Then since $e_0'=-\frac{13}{2}e_0+\frac{5}{2}e_1$ and $e_1'=-18e_0+7e_1$, we obtain the first column of $P_{e'}^{e}$ as $\begin{bmatrix}-13/2\\5/2\end{bmatrix}$ and the second column of $P_{e'}^{e}$ as $\begin{bmatrix}-18\\7\end{bmatrix}$ so that we have $P_{e'}^{e}\begin{bmatrix}-13/2 & -18\\5/2 & 7\end{bmatrix}$.

Also, $P_{e'}^{e}$ (the matrix which changes the basis e' to e, e' being the source basis and e being the target basis so that $P_{e'}^{e}$ is the matrix of $(X,(e,\epsilon))\xrightarrow{Id}(X,(e',\epsilon'))$ and the linear transformation is $(X,(e',\epsilon'))\xrightarrow{B}(X,(e,\epsilon))$ is given by $P_{e'}^{e}=\begin{bmatrix}-14 & -36\\5 & 13\end{bmatrix}$.

(One can get it using the relationship $P_{e}^{e'}=\left(P_{e'}^{e}\right)^{-1}$ since $\begin{bmatrix}-13/2 & -18\\5/2 & 7\end{bmatrix}^{-1}=\begin{bmatrix}-14 & -36\\5 & 13\end{bmatrix}$.

But it is instructive to proceed ab initio and calculate $e_1=-14e_1'+5e_2', e_2=-36e_1'+13e_2'$ directly).

7. Now suppose (e'',ϵ'') is a third coordinatization of X and $P_{e'}^{e}-b$, $P_{e''}^{e}-c$ so that $b=\left[b_i^j\right]_{n\times n}, c=\left[c_j^k\right]_{n\times n}$ are supplied by $e_i'=\sum_{j=0}^{n-1}(e_j)b_i^j,\ e_j''=\sum_{k=0}^{n-1}(e_k')c_k^j$. We find

$$\begin{aligned}e''&=\sum_{k=0}^{n-1}\left(\sum_{l=0}^{n-1}e_lb_k^l\right)c_j^k\\&=\sum_{l=0}^{n-1}e_l\left(\sum_{k=0}^{n-1}b_k^lc_j^k\right)\\&=\sum_{l=0}^{n-1}e_l(bc)_j^l\end{aligned}$$

(recall that to get the entry $(bc)_j^l$ of the product matrix bc, you have to multiply the l-th row $\left[b_0^l\ldots b_{n-i}^l\right]$ of b to the j-th column $\begin{bmatrix}c^0\\\vdots\\c_j^{n-1}\end{bmatrix}$ term by term and add up: that is, the formula for bc is given by $(bc)^l_j = \sum_{k=0}^{n-1} b^l_k c^j_k$).

Thus we find that $P^e_{e"} = bc$ which should be seen in the context of the composition

$$(X,(e,\epsilon)) \xrightarrow[P^{e'}_{e"}=c]{Id} (X,(e",\epsilon")) \xrightarrow[P^{e}_{e'}=b]{Id} (X,(e,\epsilon)) =$$
$$(X,(e",\epsilon")) \xrightarrow[P^{e}_{e"}=bc]{Id} (X,(e,\epsilon))$$

Notice that in the result $P^e_{e"} = P^e_{e'} = P^{e'}_{e"}$, e' gets erased when it occurs both as a subscript and a superscript.

8. Consider now a linear transformation $X \xrightarrow{T} X$ we recall that

$X \xrightarrow{Id} X = X \xrightarrow{\sum_{k=0}^{n-1} |e_k\rangle\langle\epsilon^k|} X$ (the 'decomposition of identity'). Then we have

$X \xrightarrow{T} X = X \xrightarrow{\sum_{k=0}^{n-1} |e_k\rangle\langle\epsilon^k|} X \xrightarrow{T} X \xrightarrow{\sum_{l=0}^{n-1} |e_l\rangle\langle\epsilon^l|} X$ and hence

$$\langle(\epsilon')^j|Te'_i\rangle = \left\langle(\epsilon')^j\left|\left[\sum_{l=0}^{n-1}|e_l\rangle\langle\epsilon^l|T\left(\sum_{k=0}^{n-1}|e_k\rangle\langle\epsilon^k|\right)(e'_i)\right]\right.\right\rangle$$
$$= \sum_{k=0}^{n-1}\sum_{l=0}^{n-1}\langle(\epsilon')^j|e_l\rangle\langle\epsilon^l|Te_k\rangle\langle\epsilon^k|e'_i\rangle$$
$$= \sum_{k=0}^{n-1}\sum_{l=0}^{n-1}(b^{-1})^j_l\langle\epsilon^l|Te_k\rangle b^k_i\rangle$$

(we used $(b')^j_l = (b^{-1})^j_l = \langle(\epsilon')^j|e_l\rangle$ and $\langle\epsilon^k|e'_i\rangle = b^k_i$ where $b = P^e_{e'}, b^{-1} = P^{e'}_e$, b being the matrix of $(X,(e",\epsilon")) \xrightarrow[P^e_{e"}=bc]{Id} (X,(e,\epsilon))$ which changes the source basis e' to the target basis e; these have been calculated earlier).

What does this above equation say? It refers to two coordinatizations, (e,ϵ) and (e',ϵ') of X and two matrices representing the linear transformation T:

$$t^l_k := \langle\epsilon^l|Te_k\rangle \text{ for } [T]^e_e$$
$$s^j_i := \langle(\epsilon')^j|Te'_i\rangle \text{ for } [T]^{e'}_{e'}$$

(Compare with $A^d_e = a = [a^j_i]; a^j_i = \langle\delta^j|Ae_i\rangle$ in $(X,(e,\epsilon)) \xrightarrow{A} (Y,(d,\delta))$

There are two interpretations of above equation.

(i) Passive There is one operator $X \xrightarrow{T} X$ and two coordinate systems (= observers), (e,ϵ) and (e',ϵ') for X. Given $x \in X$,

$$x = \sum_{i=0}^{n-1} \left| e_i' \right\rangle (x')^i,$$

$$\text{we have } Tx = \sum_{i=0}^{n-1} \left| Te_i' \right\rangle (x')^i$$

$$= \sum_{i=0}^{n-1} \sum_{j=0}^{n-1} \left| e_j' \right\rangle s_i^j (x')^i$$

$$\left(\because Te_i' = \sum_{j=0}^{n-1} e_j' \left\langle \epsilon'^j \middle| Te_i' \right\rangle = \sum_{j=0}^{n-1} e_j' s_i^j \right)$$

$$= \sum_{j=0}^{n-1} \left| e_j' \right\rangle \left(\sum_{i=0}^{n-1} s_i^j (x')^i \right),$$

so the components of Tx with reference to the coordinate system (e', ϵ') are $\sum_{i=0}^{n-1} s_i^j (x')^i$, $0 \le j \le n-1$;

$$Tx = \sum_{j=0}^{n-1} \left| e_j' \right\rangle (Tx)^j$$

$$= \sum_{j=0}^{n-1} \left| e_j' \right\rangle \left\langle \epsilon'^j \middle| Tx \right\rangle$$

$$= \sum_{j=0}^{n-1} \left| e_j' \right\rangle \left(\sum_{i=0}^{n-1} s_i^j (x')^i \right) \quad \text{where } s_i^j = \left\langle \epsilon'^j \middle| Te_i' \right\rangle$$

For the same vector $x = \sum_{i=0}^{n-1} \left| e_i \right\rangle x^i = \sum_{i=0}^{n-1} \left| e_i \right\rangle \left\langle \epsilon^i \middle| x \right\rangle$, we have

$$Tx = \sum_{i=0}^{n-1} \left| Te_i \right\rangle x^i$$

$$= \sum_{i=0}^{n-1} \sum_{j=0}^{n-1} \left| e_j \right\rangle t_i^j x^i$$

$$\left(\because Te_i = \sum_{j=0}^{n-1} e_j \left\langle \epsilon^i \middle| Te_i \right\rangle = \sum_{j=0}^{n-1} \left| e_j \right\rangle t_i^j \right)$$

$$= \sum_{j=0}^{n-1} \left| e_j \right\rangle \left(\sum_{i=0}^{n-1} t_i^j x^i \right)$$

so the components of the same vector Tx with reference to (e,ϵ) are $\sum_{i=0}^{n-1} t_i^j x^i,\ 0 \le j \le n-1;$

$$\begin{aligned} Tx &= \sum_{j=0}^{n-1} |e_j\rangle (Tx)^j \\ &= \sum_{j=0}^{n-1} |e_j\rangle \langle \epsilon^j | Tx\rangle \\ &= \sum_{j=0}^{n-1} |e_j\rangle \left(\sum_{i=0}^{n-1} t_i^j x^i \right) \text{ where } t_i^j = \langle \epsilon^j | Te_i \rangle \end{aligned}$$

Thus no two vectors moves; the two coordinate systems (= observers) measure their coordinates.

(ii) Active There is only one coordinate system but there are two operators $X \xrightarrow{T,S} X$, S:=B⁻¹TB, where B is an invertible operator on X which of course changes e_i to $Be_i = e_i'$ say. The matrix $[S]_e^e$ has entries

$$S_i^j = \langle \epsilon^j | Se_i \rangle = \sum_{k=0}^{n-1} \sum_{l=0}^{n-1} (b^{-1})_l^j t_k^l b_i^k$$

Each vector $x = \sum_{i=0}^{n-1} |e_i\rangle x^i$ moves under the action of S to

$$\begin{aligned} Sx &= \sum_{i=0}^{n-1} |Se_i\rangle x^i \\ &= \sum_{j=0}^{n-1} |e_j\rangle (Sx)^j \text{ where } (Sx)^j = \sum_{i=0}^{n-1} s_i^j x^i \\ &\left(s_i^j = \langle \epsilon^j | Se_i \rangle = \langle (\epsilon')^j | Te_i' \rangle = \sum_{k=0}^{n-1} \sum_{l=0}^{n-1} (b^{-1})_l^j t_k^l b_i^k \right) \end{aligned}$$

that is, $S = B^{-1}TB$, and the (j × i)-entry of the matrix of S is $s_i^j = \sum_{k=0}^{n-1} \sum_{l=0}^{n-1} (b^{-1})_l^j t_k^l b_i^k$, all matrices being calculated with respect to the same coordinate-system (e,ϵ)

(iii) The 'Active interpretation' prompts a definition:
Two n × n matrices S and T are called similar iff there exists an invertible n × n matrix B such that $S = B^{-1}TB$.

This is clearly an equivalence relation on $Mat_n(\mathbb{F})$.

Reflexivity: $T = (Id_n)^{-1} T\, (Id_n)$

Symmetry: $S = B^{-1}TB$ iff $T = (B^{-1})^{-1}SB^{-1}$

Transitivity: $S = B^{-1}TB$, $R = A^{-1}SA$ ensure $R = (BA)^{-1}T\,(BA)$. Now taking T to be Id and noting $Id = \sum_{i=0}^{n-1} |e_i\rangle\langle\ |$, we find that the j-th component of a vector $x \in X$ with respect to $e' = (e', \epsilon)$.

Matrix Multiplication

In mathematics, matrix multiplication or the matrix product is a binary operation that produces a matrix from two matrices. The definition is motivated by linear equations and linear transformations on vectors, which have numerous applications in applied mathematics, physics, and engineering. In more detail, if A is an $n \times m$ matrix and B is an $m \times p$ matrix, their matrix product AB is an $n \times p$ matrix, in which the m entries across a row of A are multiplied with the m entries down a columns of B and summed to produce an entry of AB. When two linear transformations are represented by matrices, then the matrix product represents the composition of the two transformations.

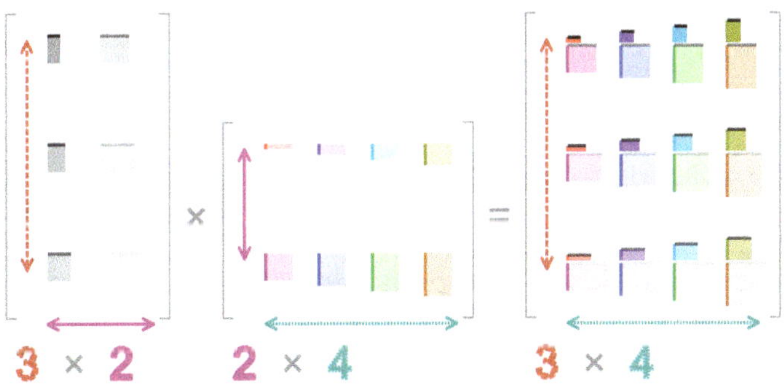

Multiplication of two matrices illustrated with a 3×2 and a 2×4 matrix with arbitrary symbols

The matrix product is not commutative in general, although it is associative and is distributive over matrix addition. The identity element of the matrix product is the identity matrix (analogous to multiplying numbers by 1), and a square matrix may have an inverse matrix (analogous to the multiplicative inverse of a number). Determinant multiplicativity applies to the matrix product. The matrix product is also important for matrix groups, and the theory of group representations and irreps.

Computing matrix products is both a central operation in many numerical algorithms and potentially time consuming, making it one of the most well-studied problems in numerical computing. Various algorithms have been devised for computing C = AB, especially for large matrices.

Notation

This article will use the following notational conventions: matrices are represented by capital letters in bold, e.g. **A**, vectors in lowercase bold, e.g. **a**, and entries of vectors and matrices are italic (since they are numbers from a field), e.g. A and a. Index notation is often the clearest way to express definitions, and is used as standard in the literature. The i, j entry of matrix **A** is indicated by $(\mathbf{A})_{ij}$ or A_{ij}, whereas a numerical label (not matrix entries) on a collection of matrices is subscripted only, e.g. $\mathbf{A}_1$, $\mathbf{A}_2$, etc.

Matrix Product (Two Matrices)

Assume two matrices are to be multiplied (the generalization to any number is discussed below).

General Definition of the Matrix Product

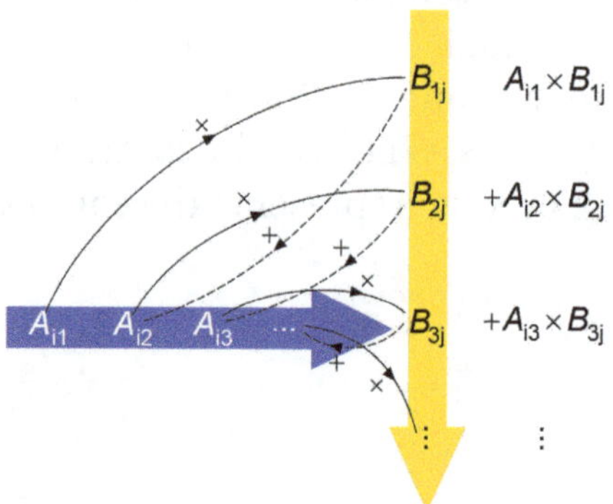

Arithmetic process of multiplying numbers (solid lines) in row i in matrix **A** and column j in matrix **B**, *then* adding the terms (dashed lines) to obtain entry ij in the final matrix.

If **A** is an $n \times m$ matrix and **B** is an $m \times p$ matrix,

$$\mathbf{A} = \begin{pmatrix} A_{11} & A_{12} & \cdots & A_{1m} \\ A_{21} & A_{22} & \cdots & A_{2m} \\ \vdots & \vdots & \ddots & \vdots \\ A_{n1} & A_{n2} & \cdots & A_{nm} \end{pmatrix}, \quad \mathbf{B} = \begin{pmatrix} B_{11} & B_{12} & \cdots & B_{1p} \\ B_{21} & B_{22} & \cdots & B_{2p} \\ \vdots & \vdots & \ddots & \vdots \\ B_{m1} & B_{m2} & \cdots & B_{mp} \end{pmatrix}$$

the matrix product **AB** (denoted without multiplication signs or dots) is defined to be the $n \times p$ matrix

$$\mathbf{AB} = \begin{pmatrix} (\mathbf{AB})_{11} & (\mathbf{AB})_{12} & \cdots & (\mathbf{AB})_{1p} \\ (\mathbf{AB})_{21} & (\mathbf{AB})_{22} & \cdots & (\mathbf{AB})_{2p} \\ \vdots & \vdots & \ddots & \vdots \\ (\mathbf{AB})_{n1} & (\mathbf{AB})_{n2} & \cdots & (\mathbf{AB})_{np} \end{pmatrix}$$

where each i, j entry is given by multiplying the entries A_{ik} (across row i of A) by the entries B_{kj} (down column j of B), for $k = 1, 2, \ldots, m$, and summing the results over k:

$$(AB)_{ij} = \sum_{k=1}^{m} A_{ik} B_{kj}.$$

Thus the product AB is defined only if the number of columns in **A** is equal to the number of rows in B, in this case m. Each entry may be computed one at a time. Sometimes, the summation convention is used as it is understood to sum over the repeated index k. To prevent any ambiguity, this convention will not be used in the article.

Usually the entries are numbers or expressions, but can even be matrices themselves. The matrix product can still be calculated exactly the same way. See below for details on how the matrix product can be calculated in terms of blocks taking the forms of rows and columns.

Illustration

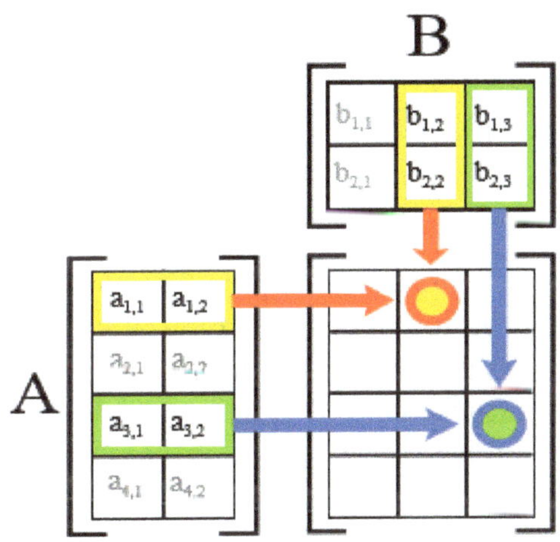

The figure to the right illustrates diagrammatically the product of two matrices A and B, showing how each intersection in the product matrix corresponds to a row of A and a column of B.

$$\overset{4\times 2 \text{ matrix}}{\begin{bmatrix} a_{11} & a_{12} \\ \cdot & \cdot \\ a_{31} & a_{32} \\ \cdot & \cdot \end{bmatrix}} \overset{2\times 3 \text{ matrix}}{\begin{bmatrix} \cdot & b_{12} & b_{13} \\ \cdot & b_{22} & b_{23} \end{bmatrix}} = \overset{4\times 3 \text{ matrix}}{\begin{bmatrix} \cdot & x_{12} & x_{13} \\ \cdot & \cdot & \cdot \\ \cdot & x_{32} & x_{33} \\ \cdot & \cdot & \cdot \end{bmatrix}}$$

The values at the intersections marked with circles are:

$$x_{12} = a_{11}b_{12} + a_{12}b_{22}$$
$$x_{33} = a_{31}b_{13} + a_{32}b_{23}$$

Examples of Matrix Products

Row Vector and Column Vector

If

$$A = \begin{pmatrix} a & b & c \end{pmatrix}, \quad B = \begin{pmatrix} x \\ y \\ z \end{pmatrix},$$

their matrix products are:

$$AB = \begin{pmatrix} a & b & c \end{pmatrix} \begin{pmatrix} x \\ y \\ z \end{pmatrix} = ax + by + cz,$$

and

$$BA = \begin{pmatrix} x \\ y \\ z \end{pmatrix} \begin{pmatrix} a & b & c \end{pmatrix} = \begin{pmatrix} xa & xb & xc \\ ya & yb & yc \\ za & zb & zc \end{pmatrix}.$$

Note AB and BA are two different matrices: the first is a 1×1 matrix while the second is a 3×3 matrix. Such expressions occur for real-valued Euclidean vectors in Cartesian coordinates, displayed as row and column matrices, in which case AB is the matrix form of their dot product, while BA the matrix form of their dyadic or tensor product.

Square Matrix and Column Vector

If

$$A = \begin{pmatrix} a & b & c \\ p & q & r \\ u & v & w \end{pmatrix}, \quad B = \begin{pmatrix} x \\ y \\ z \end{pmatrix},$$

their matrix product is:

$$AB = \begin{pmatrix} a & b & c \\ p & q & r \\ u & v & w \end{pmatrix} \begin{pmatrix} x \\ y \\ z \end{pmatrix} = \begin{pmatrix} ax + by + cz \\ px + qy + rz \\ ux + vy + wz \end{pmatrix},$$

however BA is not defined.

The product of a square matrix multiplied by a column matrix arises naturally in linear algebra; for solving linear equations and representing linear transformations. By choosing *a, b, c, p, q, r, u, v, w* in **A** appropriately, **A** can represent a variety of transformations such as rotations, scaling and reflections, shears, of a geometric shape in space.

Square Matrices

If

$$A = \begin{pmatrix} a & b & c \\ p & q & r \\ u & v & w \end{pmatrix}, \quad B = \begin{pmatrix} \alpha & \beta & \gamma \\ \lambda & \mu & \nu \\ \rho & \sigma & \tau \end{pmatrix},$$

their matrix products are:

$$AB = \begin{pmatrix} a & b & c \\ p & q & r \\ u & v & w \end{pmatrix}\begin{pmatrix} \alpha & \beta & \gamma \\ \lambda & \mu & \nu \\ \rho & \sigma & \tau \end{pmatrix} = \begin{pmatrix} a\alpha + b\lambda + c\rho & a\beta + b\mu + c\sigma & a\gamma + b\nu + c\tau \\ p\alpha + q\lambda + r\rho & p\beta + q\mu + r\sigma & p\gamma + q\nu + r\tau \\ u\alpha + v\lambda + w\rho & u\beta + v\mu + w\sigma & u\gamma + v\nu + w\tau \end{pmatrix},$$

and

$$BA = \begin{pmatrix} \alpha & \beta & \gamma \\ \lambda & \mu & \nu \\ \rho & \sigma & \tau \end{pmatrix}\begin{pmatrix} a & b & c \\ p & q & r \\ u & v & w \end{pmatrix} = \begin{pmatrix} \alpha a + \beta p + \gamma u & \alpha b + \beta q + \gamma v & \alpha c + \beta r + \gamma w \\ \lambda a + \mu p + \nu u & \lambda b + \mu q + \nu v & \lambda c + \mu r + \nu w \\ \rho a + \sigma p + \tau u & \rho b + \sigma q + \tau v & \rho c + \sigma r + \tau w \end{pmatrix}.$$

In this case, both products AB and BA are defined, and the entries show that AB and **BA** are not equal in general. Multiplying square matrices which represent linear transformations corresponds to the composite transformation.

Row Vector, Square Matrix, and Column Vector

If

$$A = \begin{pmatrix} a & b & c \end{pmatrix}, \quad B = \begin{pmatrix} \alpha & \beta & \gamma \\ \lambda & \mu & \nu \\ \rho & \sigma & \tau \end{pmatrix}, \quad C = \begin{pmatrix} x \\ y \\ z \end{pmatrix},$$

their matrix product is:

$$\mathrm{ABC} = \begin{pmatrix} a & b & c \end{pmatrix}\left[\begin{pmatrix} \alpha & \beta & \gamma \\ \lambda & \mu & \nu \\ \rho & \sigma & \tau \end{pmatrix}\begin{pmatrix} x \\ y \\ z \end{pmatrix}\right] = \left[\begin{pmatrix} a & b & c \end{pmatrix}\begin{pmatrix} \alpha & \beta & \gamma \\ \lambda & \mu & \nu \\ \rho & \sigma & \tau \end{pmatrix}\right]\begin{pmatrix} x \\ y \\ z \end{pmatrix}$$

$$= \begin{pmatrix} a & b & c \end{pmatrix}\begin{pmatrix} \alpha x + \beta y + \gamma z \\ \lambda x + \mu y + \nu z \\ \rho x + \sigma y + \tau z \end{pmatrix} = \begin{pmatrix} a\alpha + b\lambda + c\rho & a\beta + b\mu + c\sigma & a\gamma + b\nu + c\tau \end{pmatrix}\begin{pmatrix} x \\ y \\ z \end{pmatrix}$$

$$= a\alpha x + b\lambda x + c\rho x + a\beta y + b\mu y + c\sigma y + a\gamma z + b\nu z + c\tau z,$$

however CBA is not defined. Note that A(BC) = (AB)C, this is one of many general properties listed below. Expressions of the form ABC occur when calculating the inner product of two vectors displayed as row and column vectors in an arbitrary coordinate system, and the metric tensor in these coordinates written as the square matrix.

Rectangular Matrices

If

$$\mathrm{A} = \begin{pmatrix} a & b & c \\ x & y & z \end{pmatrix}, \quad \mathrm{B} = \begin{pmatrix} \alpha & \rho \\ \beta & \sigma \\ \gamma & \tau \end{pmatrix},$$

their matrix products are:

$$\mathrm{AB} = \begin{pmatrix} a & b & c \\ x & y & z \end{pmatrix}\begin{pmatrix} \alpha & \rho \\ \beta & \sigma \\ \gamma & \tau \end{pmatrix} = \begin{pmatrix} a\alpha + b\beta + c\gamma & a\rho + b\sigma + c\tau \\ x\alpha + y\beta + z\gamma & x\rho + y\sigma + z\tau \end{pmatrix},$$

and

$$\mathrm{BA} = \begin{pmatrix} \alpha & \rho \\ \beta & \sigma \\ \gamma & \tau \end{pmatrix}\begin{pmatrix} a & b & c \\ x & y & z \end{pmatrix} = \begin{pmatrix} \alpha a + \rho x & \alpha b + \rho y & \alpha c + \rho z \\ \beta a + \sigma x & \beta b + \sigma y & \beta c + \sigma z \\ \gamma a + \tau x & \gamma b + \tau y & \gamma c + \tau z \end{pmatrix}.$$

Properties of the Matrix Product (Two Matrices)

Analogous to numbers (elements of a field), matrices satisfy the following general properties, although there is one subtlety, due to the nature of matrix multiplication.

All Matrices

1. Not commutative:

   In general:

$$AB \neq BA$$

   because AB and BA may not be simultaneously defined, and even if they are they may still not be equal. This is contrary to ordinary multiplication of numbers. To specify the ordering of matrix multiplication in words; "pre-multiply (or left multiply) A by B" means BA, while "post-multiply (or right multiply) A by C" means AC. As long as the entries of the matrix come from a ring that has an identity, and $n > 1$ there is a pair of $n \times n$ noncommuting matrices over the ring. A notable exception is that the identity matrix (or any scalar multiple of it) commutes with every square matrix.

   In index notation:

$$\sum_k A_{ik} B_{kj} \neq \sum_k B_{ik} A_{kj}$$

2. Distributive over matrix addition:

   Left distributivity:

$$A(B+C) = AB + AC$$

   Right distributivity:

$$(A+B)C = AC + BC$$

   In index notation, these are respectively:

$$\sum_k A_{ik}(B_{kj} + C_{kj}) = \sum_k A_{ik} B_{kj} + \sum_k A_{ik} C_{kj}$$

$$\sum_k (A_{ik} + B_{ik}) C_{kj} = \sum_k A_{ik} C_{kj} + \sum_k B_{ik} C_{kj}$$

3. Scalar multiplication is compatible with matrix multiplication:

$$\lambda(AB) = (\lambda A)B \text{ and } (AB)\lambda = A(B\lambda)$$

   where λ is a scalar. If the entries of the matrix are real or complex numbers (or from any other commutative ring), then all four quantities are equal. More generally, all four are equal if λ belongs to the center of the ring of entries of the matrix, because in this case $\lambda X = X\lambda$ for all matrices X. In index notation, these are respectively:

$$\lambda\sum_k (A_{ik}B_{kj}) = \sum_k (\lambda A_{ik})B_{kj} = \sum_k A_{ik}(\lambda B_{kj})$$

$$\sum_k (A_{ik}B_{kj})\lambda = \sum_k (A_{ik}\lambda)B_{kj} = \sum_k A_{ik}(B_{kj}\lambda)$$

4. Transpose:

$$(AB)^T = B^T A^T$$

where ^{T} denotes the transpose, the interchange of row i with column i in a matrix. This identity holds for any matrices over a commutative ring, but not for all rings in general. Note that A and B are reversed.

In index notation:

$$\begin{aligned}\left[(AB)^T\right]_{ij} &= (AB)_{ji} \\ &= \sum_k (A)_{jk}(B)_{ki} \\ &= \sum_k \left(A^T\right)_{kj}\left(B^T\right)_{ik} \\ &= \sum_k \left(B^T\right)_{ik}\left(A^T\right)_{kj} \\ &= \left[\left(B^T\right)\left(A^T\right)\right]_{ij}\end{aligned}$$

5. Complex conjugate:

If A and B have complex entries, then

$$(AB)^* = A^*B^*$$

where * denotes the entry-wise complex conjugate of a matrix.

In index notation:

$$\begin{aligned}\left[(AB)^*\right]_{ij} &= \left[\sum_k (A)_{ik}(B)_{kj}\right]^* \\ &= \sum_k (A)^*_{ik}(B)^*_{kj} \\ &= \sum_k \left(A^*\right)_{ik}\left(B^*\right)_{kj} \\ &= \left(A^*B^*\right)_{ij}\end{aligned}$$

6. Conjugate transpose:

   If A and B have complex entries, then

$$(AB)^{\dagger} = B^{\dagger}A^{\dagger}$$

   where † denotes the Conjugate transpose of a matrix (complex conjugate and transposed).

   In index notation:

$$\left[(AB)^{\dagger}\right]_{ij} = \left[(AB)^{*}\right]_{ji}$$
$$= \sum_{k}\left(A^{*}\right)_{jk}\left(B^{*}\right)_{ki}$$
$$= \sum_{k}\left(A^{\dagger}\right)_{kj}\left(B^{\dagger}\right)_{ik}$$
$$= \sum_{k}\left(B^{\dagger}\right)_{ik}\left(A^{\dagger}\right)_{kj}$$
$$= \left[\left(B^{\dagger}\right)\left(A^{\dagger}\right)\right]_{ij}$$

7. Traces:

   The trace of a product AB is independent of the order of A and B:

$$\mathrm{tr}(AB) \quad \mathrm{tr}(BA)$$

   In index notation:

$$\mathrm{tr}(AB) = \sum_{i}\sum_{k} A_{ik}B_{ki}$$
$$= \sum_{k}\sum_{i} B_{ki}A_{ik}$$
$$= \mathrm{tr}(BA)$$

Square Matrices Only

1. Identity element:

   If A is a square matrix, then

$$AI = IA = A$$

   where I is the identity matrix of the same order.

2. Inverse matrix:

   If A is a square matrix, there may be an inverse matrix A^{-1} of A such that

$$AA^{-1} = A^{-1}A = I$$

   If this property holds then A is an invertible matrix, if not A is a singular matrix. Moreover,

$$(AB)^{-1} = B^{-1}A^{-1}$$

3. Determinants:

   When a determinant of a matrix is defined (i.e., when the underlying ring is commutative), if A and B are square matrices of the same order, the determinant of their product AB equals the product of their determinants:

$$\det(AB) = \det(A)\det(B)$$

   Since det(A) and det(B) are elements of the said commutative ring, det(A)det(B) = det(B)det(A), and so det(AB) = det(BA), even when AB ≠ BA.

Matrix Product (Any Number)

Matrix multiplication can be extended to the case of more than two matrices, provided that for each sequential pair, their dimensions match.

The product of n matrices $A_1, A_2, \ldots, A_n$ with sizes $s_0 \times s_1, s_1 \times s_2, \ldots, s_{n-1} \times s_n$ (where $s_0, s_1, s_2, \ldots, s_n$ are all simply positive integers and the subscripts are labels corresponding to the matrices, nothing more), is the $s_0 \times s_n$ matrix:

$$\prod_{i=1}^{n} A_i = A_1 A_2 \cdots A_n.$$

In index notation:

$$(A_1 A_2 \cdots A_n)_{i_0 i_n} = \sum_{i_1=1}^{s_1} \sum_{i_2=1}^{s_2} \cdots \sum_{i_{n-1}=1}^{s_{n-1}} (A_1)_{i_0 i_1} (A_2)_{i_1 i_2} (A_3)_{i_2 i_3} \cdots (A_{n-1})_{i_{n-2} i_{n-1}} (A_n)_{i_{n-1} i_n}$$

Properties of the Matrix Product (Any Number)

The same properties will hold, as long as the ordering of matrices is not changed. Some of the previous properties for more than two matrices generalize as follows.

1. Associative:

   The matrix product is associative. If three matrices A, B, and C are respectively $m \times p$, $p \times q$, and $q \times r$ matrices, then there are two ways of grouping them without changing their order, and

$$ABC = A(BC) = (AB)C$$

   is an $m \times r$ matrix.

   If four matrices A, B, C, and D are respectively $m \times p$, $p \times q$, $q \times r$, and $r \times s$ matrices, then there are five ways of grouping them without changing their order, and

$$ABCD = ((AB)C)D = (A(BC))D = A((BC)D) = A(B(CD)) = (AB)(CD)$$

   is an $m \times s$ matrix.

   In general, the number of possible ways of grouping n matrices for multiplication is equal to the $(n - 1)$th Catalan number

2. Trace:

   The trace of a product of n matrices $A_1, A_2, ..., A_n$ is invariant under cyclic permutations of the matrices in the product:

$$tr(A_1A_2A_3 \ldots A_{n-2}A_{n-1}A_n) = tr(A_2A_3A_4 \ldots A_{n-1}A_nA_1) = tr(A_3A_4A_5 \ldots A_nA_1A_2) = \ldots$$

3. Determinant:

   For square matrices only, the determinant of a product is the product of determinants:

$$\det\left(\prod_{i=1}^{n} A_i\right) = \prod_{i=1}^{n} \det(A_i)$$

Examples of Chain Multiplication

Similarity transformations involving similar matrices are matrix products of the three square matrices, in the form:

$$B = P^{-1}AP$$

where P is the similarity matrix and A and B are said to be similar if this relation holds. This product appears frequently in linear algebra and applications, such as diagonalizing square matrices and the equivalence between different matrix representations of the same linear operator.

Operations Derived from the Matrix Product

More operations on square matrices can be defined using the matrix product, such as powers and nth roots by repeated matrix products, the matrix exponential can be defined by a power series, the matrix logarithm is the inverse of matrix exponentiation, and so on.

Powers of Matrices

Square matrices can be multiplied by themselves repeatedly in the same way as ordinary numbers, because they always have the same number of rows and columns. This repeated multiplication can be described as a power of the matrix, a special case of the ordinary matrix product. On the contrary, *rectangular* matrices do not have the same number of rows and columns so they can *never* be raised to a power. An $n \times n$ matrix **A** raised to a positive integer k is defined as

$$A^k = \underbrace{AA\cdots A}_{k\,\text{times}}$$

and the following identities hold, where λ is a scalar:

1. Zero power:

$$A^0 = I$$

   where **I** is the identity matrix. This is parallel to the zeroth power of any number which equals unity.

2. Scalar multiplication:

$$(\lambda A)^k = \lambda^k A^k$$

3. Determinant:

$$\det(A^k) = \det(A)^k$$

The naive computation of matrix powers is to multiply k times the matrix A to the result, starting with the identity matrix just like the scalar case. This can be improved using exponentiation by squaring, a method commonly used for scalars. For diagonalizable matrices, an even better method is to use the eigenvalue decomposition of A. Another method based on the Cayley–Hamilton theorem finds an identity using the matrices' characteristic polynomial, producing a more effective equation for A^k in which a *scalar* is raised to the required power, rather than an entire *matrix*.

A special case is the power of a diagonal matrix. Since the product of diagonal matrices amounts to simply multiplying corresponding diagonal elements together, the power k of a diagonal matrix A will have entries raised to the power. Explicitly;

$$A^k = \begin{pmatrix} A_{11} & 0 & \cdots & 0 \\ 0 & A_{22} & \cdots & 0 \\ \vdots & \vdots & \ddots & \vdots \\ 0 & 0 & \cdots & A_{nn} \end{pmatrix}^k = \begin{pmatrix} A_{11}^k & 0 & \cdots & 0 \\ 0 & A_{22}^k & \cdots & 0 \\ \vdots & \vdots & \ddots & \vdots \\ 0 & 0 & \cdots & A_{nn}^k \end{pmatrix}$$

meaning it is easy to raise a diagonal matrix to a power. When raising an arbitrary matrix (not necessarily a diagonal matrix) to a power, it is often helpful to exploit this property by diagonalizing the matrix first.

Applications of the Matrix Product

Linear Transformations

Matrices offer a concise way of representing linear transformations between vector spaces, and matrix multiplication corresponds to the composition of linear transformations. The matrix product of two matrices can be defined when their entries belong to the same ring, and hence can be added and multiplied.

Let U, V, and W be vector spaces over the same field with given bases, S: $V \to W$ and T: $U \to V$ be linear transformations and ST: $U \to W$ be their composition.

Suppose that A, B, and C are the matrices representing the transformations S, T, and ST with respect to the given bases.

Then AB = C, that is, the matrix of the composition (or the product) of linear transformations is the product of their matrices with respect to the given bases.

Linear Systems of Equations

A system of linear equations with the same number of equations as variables can be solved by collecting the coefficients of the equations into a square matrix, then inverting the matrix equation.

A similar procedure can be used to solve a system of linear differential equations.

Group Theory and Representation Theory

The Inner and Outer Products

Given two *column vectors* a and b, the Euclidean inner product and outer product are the simplest special cases of the matrix product.

Inner Product

The inner product of two vectors in matrix form is equivalent to a column vector multiplied on its left by a row vector:

$$\begin{aligned} a \cdot b &= a^T b \\ &= \begin{pmatrix} a_1 & a_2 & \cdots & a_n \end{pmatrix} \begin{pmatrix} b_1 \\ b_2 \\ \vdots \\ b_n \end{pmatrix} \\ &= a_1 b_1 + a_2 b_2 + \cdots + a_n b_n \\ &= \sum_{i=1}^{n} a_i b_i, \end{aligned}$$

where a^T denotes the transpose of a.

The matrix product itself can be expressed in terms of inner product. Suppose that the first $n \times m$ matrix A is decomposed into its row vectors $\mathbf{a}_i$, and the second $m \times p$ matrix B into its column vectors b_i:

$$A = \begin{pmatrix} A_{11} & A_{12} & \cdots & A_{1m} \\ A_{21} & A_{22} & \cdots & A_{2m} \\ \vdots & \vdots & \ddots & \vdots \\ A_{n1} & A_{n2} & \cdots & A_{nm} \end{pmatrix} = \begin{pmatrix} a_1 \\ a_2 \\ \vdots \\ a_n \end{pmatrix},$$

$$B = \begin{pmatrix} B_{11} & B_{12} & \cdots & B_{1p} \\ B_{21} & B_{22} & \cdots & B_{2p} \\ \vdots & \vdots & \ddots & \vdots \\ B_{m1} & B_{m2} & \cdots & B_{mp} \end{pmatrix} = \begin{pmatrix} b_1 & b_2 & \cdots & b_p \end{pmatrix}$$

where

$$a_i = \begin{pmatrix} A_{i1} & A_{i2} & \cdots & A_{im} \end{pmatrix}, \quad b_i = \begin{pmatrix} B_{1i} \\ B_{2i} \\ \vdots \\ B_{mi} \end{pmatrix}$$

Then:

$$AB = \begin{pmatrix} a_1 \\ a_2 \\ \vdots \\ a_n \end{pmatrix} \begin{pmatrix} b_1 & b_2 & \ldots & b_p \end{pmatrix} = \begin{pmatrix} (a_1 \cdot b_1) & (a_1 \cdot b_2) & \ldots & (a_1 \cdot b_p) \\ (a_2 \cdot b_1) & (a_2 \cdot b_2) & \ldots & (a_2 \cdot b_p) \\ \vdots & \vdots & \ddots & \vdots \\ (a_n \cdot b_1) & (a_n \cdot b_2) & \ldots & (a_n \cdot b_p) \end{pmatrix}$$

It is also possible to express a matrix product in terms of concatenations of products of matrices and row or column vectors:

$$AB = \begin{pmatrix} Ab_1 & Ab_2 & \dots & Ab_p \end{pmatrix} = \begin{pmatrix} a_1B \\ a_2B \\ \vdots \\ a_nB \end{pmatrix}$$

These decompositions are particularly useful for matrices that are envisioned as concatenations of particular types of row vectors or column vectors, e.g. orthogonal matrices (whose rows and columns are unit vectors orthogonal to each other) and Markov matrices (whose rows or columns sum to 1).

Outer Product

The outer product (also known as the dyadic product or tensor product) of two vectors in matrix form is equivalent to a row vector multiplied on the left by a column vector:

$$\begin{aligned} a \otimes b &= ab^T \\ &= \begin{pmatrix} a_1 \\ a_2 \\ \vdots \\ a_n \end{pmatrix} \begin{pmatrix} b_1 & b_2 & \cdots & b_n \end{pmatrix} \\ &= \begin{pmatrix} a_1b_1 & a_1b_2 & \cdots & a_1b_n \\ a_2b_1 & a_2b_2 & \cdots & a_2b_n \\ \vdots & \vdots & \ddots & \vdots \\ a_nb_1 & a_nb_2 & \cdots & a_nb_n \end{pmatrix}. \end{aligned}$$

An alternative method is to express the matrix product in terms of the outer product. The decomposition is done the other way around, the first matrix A is decomposed into column vectors $\overline{a}_i$ and the second matrix B into row vectors $\overline{b}_i$:

$$\begin{aligned} AB &= \begin{pmatrix} \overline{a}_1 & \overline{a}_2 & \cdots & \overline{a}_m \end{pmatrix} \begin{pmatrix} \overline{b}_1 \\ \overline{b}_2 \\ \vdots \\ \overline{b}_m \end{pmatrix} \\ &= \overline{a}_1 \otimes \overline{b}_1 + \overline{a}_2 \otimes \overline{b}_2 + \cdots + \overline{a}_m \otimes \overline{b}_m \\ &= \sum_{i=1}^{m} \overline{a}_i \otimes \overline{b}_i \end{aligned}$$

where this time

$$\bar{a}_i = \begin{pmatrix} A_{1i} \\ A_{2i} \\ \vdots \\ A_{ni} \end{pmatrix}, \quad \bar{b}_i = \begin{pmatrix} B_{i1} & B_{i2} & \cdots & B_{ip} \end{pmatrix}.$$

This method emphasizes the effect of individual column/row pairs on the result, which is a useful point of view with e.g. covariance matrices, where each such pair corresponds to the effect of a single sample point.

$$\begin{pmatrix} 1 & 2 & 3 \\ 4 & 5 & 6 \\ 7 & 8 & 9 \end{pmatrix}\begin{pmatrix} a & d \\ b & c \\ c & f \end{pmatrix} = \begin{pmatrix} 1 \\ 4 \\ 7 \end{pmatrix} \otimes \begin{pmatrix} a & d \end{pmatrix} + \begin{pmatrix} 2 \\ 5 \\ 8 \end{pmatrix} \otimes \begin{pmatrix} b & e \end{pmatrix} + \begin{pmatrix} 3 \\ 6 \\ 9 \end{pmatrix} \otimes \begin{pmatrix} c & f \end{pmatrix}$$

$$= \begin{pmatrix} 1a & 1d \\ 4a & 4d \\ 7a & 7d \end{pmatrix} + \begin{pmatrix} 2b & 2e \\ 5b & 5e \\ 8b & 8e \end{pmatrix} + \begin{pmatrix} 3c & 3f \\ 6c & 6f \\ 9c & 9f \end{pmatrix}$$

$$= \begin{pmatrix} 1a + 2b + 3c & 1d + 2e + 3f \\ 4a + 5b + 6c & 4d + 5e + 6f \\ 7a + 8b + 9c & 7d + 8e + 9f \end{pmatrix}.$$

Algorithms for Efficient Matrix Multiplication

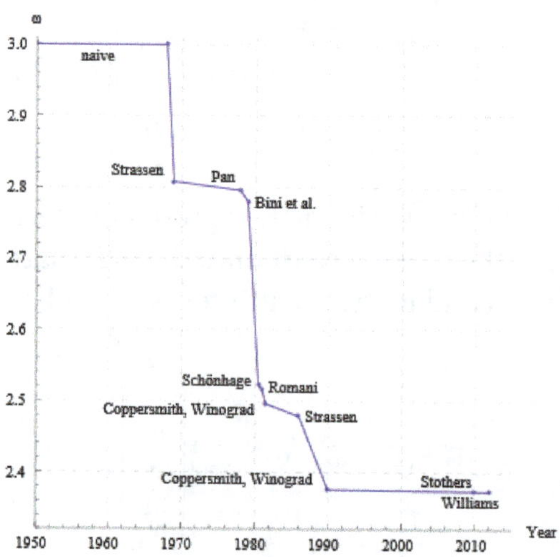

The bound on ω over time.

The running time of square matrix multiplication, if carried out naïvely, is $O(n^3)$. The running time for multiplying rectangular matrices (one $m \times p$-matrix with one $p \times$

n-matrix) is $O(mnp)$, however, more efficient algorithms exist, such as Strassen's algorithm, devised by Volker Strassen in 1969 and often referred to as "fast matrix multiplication". It is based on a way of multiplying two 2 × 2-matrices which requires only 7 multiplications (instead of the usual 8), at the expense of several additional addition and subtraction operations. Applying this recursively gives an algorithm with a multiplicative cost of $O(n^{\log_2 7}) \approx O(n^{2.807})$. Strassen's algorithm is more complex, and the numerical stability is reduced compared to the naïve algorithm. Nevertheless, it appears in several libraries, such as BLAS, where it is significantly more efficient for matrices with dimensions $n > 100$, and is very useful for large matrices over exact domains such as finite fields, where numerical stability is not an issue.

The current $O(n^k)$ algorithm with the lowest known exponent k is a generalization of the Coppersmith–Winograd algorithm that has an asymptotic complexity of $O(n^{2.3728639})$, by François Le Gall. This algorithm, and the Coppersmith–Winograd algorithm on which it is based, are similar to Strassen's algorithm: a way is devised for multiplying two $k \times k$-matrices with fewer than k^3 multiplications, and this technique is applied recursively. However, the constant coefficient hidden by the Big O notation is so large that these algorithms are only worthwhile for matrices that are too large to handle on present-day computers.

Since any algorithm for multiplying two $n \times n$-matrices has to process all $2 \times n^2$-entries, there is an asymptotic lower bound of $\Omega(n^2)$ operations. Raz (2002) proves a lower bound of $\Omega(n^2 \log(n))$ for bounded coefficient arithmetic circuits over the real or complex numbers.

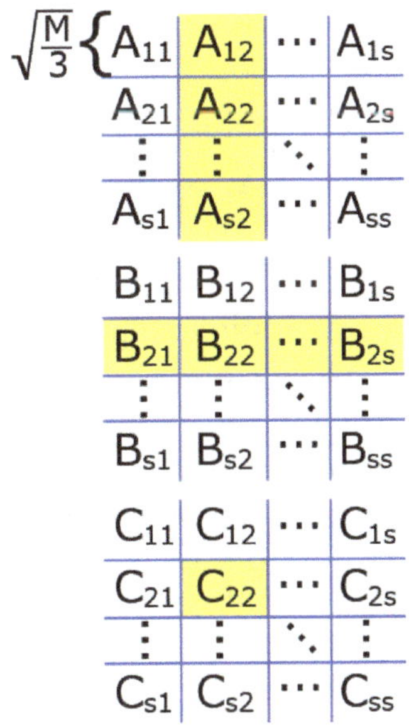

Block matrix multiplication. In the 2D algorithm, each processor is responsible for one submatrix of C. In the 3D algorithm, every pair of submatrices from A and B that is multiplied is assigned to one processor.

Cohn *et al.* (2003, 2005) put methods such as the Strassen and Coppersmith–Winograd algorithms in an entirely different group-theoretic context, by utilising triples of subsets of finite groups which satisfy a disjointness property called the triple product property (TPP). They show that if families of wreath products of Abelian groups with symmetric groups realise families of subset triples with a simultaneous version of the TPP, then there are matrix multiplication algorithms with essentially quadratic complexity. Most researchers believe that this is indeed the case. However, Alon, Shpilka and Chris Umans have recently shown that some of these conjectures implying fast matrix multiplication are incompatible with another plausible conjecture, the sunflower conjecture.

Freivalds' algorithm is a simple Monte Carlo algorithm that given matrices A, B, C verifies in $\Theta(n^2)$ time if AB = C.

Parallel Matrix Multiplication

Because of the nature of matrix operations and the layout of matrices in memory, it is typically possible to gain substantial performance gains through use of parallelization and vectorization. Several algorithms are possible, among which divide and conquer algorithms based on the block matrix decomposition

$$C = \begin{pmatrix} C_{11} & C_{12} \\ C_{21} & C_{22} \end{pmatrix} = \begin{pmatrix} A_{11} & A_{12} \\ A_{21} & A_{22} \end{pmatrix} \begin{pmatrix} B_{11} & B_{12} \\ B_{21} & B_{22} \end{pmatrix} = AB$$

that also underlies Strassen's algorithm. Here, A, B and **C** are presumed to be n by n (square) matrices, and C_{11} etc. are $n/2$ by $n/2$ submatrices. From this decomposition, one derives

$$\begin{pmatrix} A_{11} & A_{12} \\ A_{21} & A_{22} \end{pmatrix} \begin{pmatrix} B_{11} & B_{12} \\ B_{21} & B_{22} \end{pmatrix} = \begin{pmatrix} A_{11}B_{11} + A_{12}B_{21} & A_{11}B_{12} + A_{12}B_{22} \\ A_{21}B_{11} + A_{22}B_{21} & A_{21}B_{12} + A_{22}B_{22} \end{pmatrix}$$

which consists of eight multiplications of pairs of submatrices, which can all be performed in parallel, followed by an addition step. Applying this recursively, and performing the additions in parallel as well, one obtains an algorithm that runs in $\Theta(\log^2 n)$ time on an ideal machine with an infinite number of processors, and has a maximum possible speedup of $\Theta(n^3/(\log^2 n))$ on any real computer (although the algorithm isn't practical, a more practical variant achieves $\Theta(n^2)$ speedup).

It should be noted that some lower time-complexity algorithms on paper may have indirect time complexity costs on real machines.

Communication-avoiding and Distributed Algorithms

On modern architectures with hierarchical memory, the cost of loading and storing input matrix elements tends to dominate the cost of arithmetic. On a single machine this is the amount of data transferred between RAM and cache, while on a distributed memory multi-node machine it is the amount transferred between nodes; in either case it is called the *communication bandwidth*. The naïve algorithm using three nested loops uses $\Omega(n^3)$ communication bandwidth.

Cannon's algorithm, also known as the *2D algorithm*, partitions each input matrix into a block matrix whose elements are submatrices of size $\sqrt{M/3}$ by $\sqrt{M/3}$, where M is the size of fast memory. The naïve algorithm is then used over the block matrices, computing products of submatrices entirely in fast memory. This reduces communication bandwidth to $O(n^3/\sqrt{M})$, which is asymptotically optimal (for algorithms performing $\Omega(n^3)$ computation).

In a distributed setting with p processors arranged in a $\sqrt{p}$ by $\sqrt{p}$ 2D mesh, one submatrix of the result can be assigned to each processor, and the product can be computed with each processor transmitting $O(n^2/\sqrt{p})$ words, which is asymptotically optimal assuming that each node stores the minimum $O(n^2/p)$ elements. This can be improved by the *3D algorithm,* which arranges the processors in a 3D cube mesh, assigning every product of two input submatrices to a single processor. The result submatrices are then generated by performing a reduction over each row. This algorithm transmits $O(n^2/p^{2/3})$ words per processor, which is asymptotically optimal. However, this requires replicating each input matrix element $p^{1/3}$ times, and so requires a factor of $p^{1/3}$ more memory than is needed to store the inputs. This algorithm can be combined with Strassen to further reduce runtime. "2.5D" algorithms provide a continuous tradeoff between memory usage and communication bandwidth. On modern distributed computing environments such as MapReduce, specialized multiplication algorithms have been developed.

Other Forms of Multiplication

The term "matrix multiplication" is most commonly reserved for the definition given in this article. It could be more loosely applied to other definitions.

- Hadamard product (matrices)
- Frobenius inner product
- Kronecker product
- Cracovian product

Matrix Decomposition

In the mathematical discipline of linear algebra, a matrix decomposition or matrix factorization is a factorization of a matrix into a product of matrices. There are many different matrix decompositions; each finds use among a particular class of problems.

Example

In numerical analysis, different decompositions are used to implement efficient matrix algorithms.

For instance, when solving a system of linear equations $Ax = b$, the matrix *A* can be decomposed via the LU decomposition. The LU decomposition factorizes a matrix into a lower triangular matrix *L* and an upper triangular matrix *U*. The systems $L(Ux) = b$ and $Ux = L^{-1}b$ require fewer additions and multiplications to solve, compared with the original system $Ax = b$, , though one might require significantly more digits in inexact arithmetic such as floating point.

Similarly, the QR decomposition expresses *A* as *QR* with *Q* an orthogonal matrix and *R* an upper triangular matrix. Thc system $Q(Rx) = b$ is solved by $Rx = Q^Tb = c$, and the system $Rx = c$ is solved by 'back substitution'. The number of additions and multiplications required is about twice that of using the LU solver, but no more digits are required in inexact arithmetic because the QR decomposition is numerically stable.

Decompositions Related to Solving Systems of Linear Equations

- Applicable to: square matrix *A*
- Decomposition: $A = LU$, where *L* is lower triangular and *U* is upper triangular
- Related: the *LDU* decomposition is $A = LDU$, where *L* is lower triangular with ones on the diagonal, *U* is upper triangular with ones on the diagonal, and *D* is a diagonal matrix.
- Related: the *LUP* decomposition is $A = LUP$, where *L* is lower triangular, *U* is upper triangular, and *P* is a permutation matrix.
- Existence: An LUP decomposition exists for any square matrix *A*. When *P* is an identity matrix, the LUP decomposition reduces to the LU decomposition. If the LU decomposition exists, then the LDU decomposition exists.
- Comments: The LUP and LU decompositions are useful in solving an *n*-by-*n* system of linear equations $Ax = b$. These decompositions summarize the process of Gaussian elimination in matrix form. Matrix *P* represents any row interchanges carried out in the process of Gaussian elimination. If Gaussian elimination produces the row echelon form without requiring any row interchanges, then $P = I$, so an LU decomposition exists.

LU Reduction

Rank Factorization

- Applicable to: *m*-by-*n* matrix *A* of rank *r*

- Decomposition: $A = CF$ where C is an m-by-r full column rank matrix and F is an r-by-n full row rank matrix
- Comment: The rank factorization can be used to compute the Moore–Penrose pseudoinverse of A, which one can apply to obtain all solutions of the linear system $Ax = b$.

Cholesky Decomposition

- Applicable to: square, hermitian, positive definite matrix A
- Decomposition: $A = U^*U$, where U is upper triangular with real positive diagonal entries
- Comment: if the matrix **A** is Hermitian and positive semi-definite, then it has a decomposition of the form $A = U^*U$ if the diagonal entries of U are allowed to be zero
- Uniqueness: for positive definite matrices Cholesky decomposition is unique. However, it is not unique in the positive semi-definite case.
- Comment: if A is real and symmetric, U has all real elements
- Comment: An alternative is the LDL decomposition, which can avoid extracting square roots.

QR Decomposition

- Applicable to: m-by-n matrix A
- Decomposition: $A = QR$ where Q is an unitary matrix of size m-by-m, and R is an upper triangular matrix of size m-by-n
- Uniqueness: In general it is not unique, but if A is of full rank, then there exists a single R that has all positive diagonal elements. If A is square, also Q is unique.
- Comment: The QR decomposition provides an alternative way of solving the system of equations $Ax = b$ without inverting the matrix A. The fact that Q is orthogonal means that $Q^TQ = I$, so that $Ax = b$ is equivalent to $Rx = Q^Tb$, which is easier to solve since R is triangular.

Decompositions Based on Eigenvalues and Related Concepts

Eigendecomposition

- Also called *spectral decomposition*

- Applicable to: square matrix *A* with linearly independent eigenvectors (not necessarily distinct eigenvalues).
- Decomposition: $A = VDV^{-1}$, where *D* is a diagonal matrix formed from the eigenvalues of *A*, and the columns of *V* are the corresponding eigenvectors of *A*.
- Existence: An *n*-by-*n* matrix *A* always has *n* (complex) eigenvalues, which can be ordered (in more than one way) to form an *n*-by-*n* diagonal matrix *D* and a corresponding matrix of nonzero columns *V* that satisfies the eigenvalue equation $AV = VD$. V is invertible if and only if the *n* eigenvectors are linearly independent (i.e., each eigenvalue has geometric multiplicity equal to its algebraic multiplicity). A sufficient (but not necessary) condition for this to happen is that all the eigenvalues are different (in this case geometric and algebraic multiplicity are equal to 1)
- Comment: One can always normalize the eigenvectors to have length one.
- Comment: Every normal matrix *A* (i.e., matrix for which $AA^* = A^*A$, where A^*, is a conjugate transpose) can be eigendecomposed. For normal matrix *A* (and only for normal matrix), the eigenvectors can also be made orthonormal ($VV^* = I$) and eigendecomposition reads as $A = VDV^*$. In particular all unitary, Hermitian, and skew-Hermitian (in real-valued case, all orthogonal, symmetric, and skew-symmetric, respectively) matrices are normal and therefore possess this property.
- Comment: For real symmetric matrix *A* the eigendecomposition always exists and can be written as $A = VDV^T$, where both *D* and *V* are real-valued.
- Comment: The eigendecomposition is useful for understanding the solution of a system of linear ordinary differential equations or linear difference equations. For example, the difference equation $x_{t+1} = Ax_t$ starting from the initial condition $x_0 = c$ is solved by $x_t = A^t c$, which is equivalent to $x_t = VD^tV^{-1}c$, where *V* and *D* are the matrices formed from the eigenvectors and eigenvalues of *A*. Since *D* is diagonal, raising it to power D^t, just involves raising each element on the diagonal to the power *t*. This is much easier to do and to understand than raising *A* to power *t*, since *A* is usually not diagonal.

Jordan Decomposition

The Jordan normal form and the Jordan–Chevalley decomposition

- Applicable to: square matrix *A*
- Comment: the Jordan normal form generalizes the eigendecomposition to cases where there are repeated eigenvalues and cannot be diagonalized, the Jordan–Chevalley decomposition does this without choosing a basis.

Schur Decomposition

- Applicable to: square matrix A
- Decomposition (complex version): $A = UTU^*$, where U is a unitary matrix, U^* is the conjugate transpose of U, and T is an upper triangular matrix called the complex Schur form which has the eigenvalues of A along its diagonal.
- Comment: if A is a normal matrix, then T is diagonal and the Schur decomposition coincides with the spectral decomposition.

Real Schur Decomposition

- Applicable to: square matrix A
- Decomposition: This is a version of Shur decomposition where V and S only contain real numbers. One can always write $A = VSV^T$ where V is a real orthogonal matrix, V^T is the transpose of V, and S is a block upper triangular matrix called the real Schur form. The blocks on the diagonal of S are of size 1×1 (in which case they represent real eigenvalues) or 2×2 (in which case they are derived from complex conjugate eigenvalue pairs).

QZ Decomposition

- Also called: *generalized Schur decomposition*
- Applicable to: square matrices A and B
- Comment: there are two versions of this decomposition: complex and real.
- Decomposition (complex version): $A = QSZ^H$ and $B = QTZ^H$ where Q and Z are unitary matrices, the H superscript represents conjugate transpose, and S and T are upper triangular matrices.
- Comment: in the complex QZ decomposition, the ratios of the diagonal elements of S to the corresponding diagonal elements of T, $\lambda_i = S_{ii} / T_{ii}$, are the generalized eigenvalues that solve the generalized eigenvalue problem $Av = \lambda Bv$ (where λ is an unknown scalar and v is an unknown nonzero vector).
- Decomposition (real version): $A = QSZ^T$ and $B = QTZ^T$ where A, B, Q, Z, S, and T are matrices containing real numbers only. In this case Q and Z are orthogonal matrices, the T superscript represents transposition, and S and T are block upper triangular matrices. The blocks on the diagonal of S and T are of size 1×1 or 2×2.

Takagi's Factorization

- Applicable to: square, complex, symmetric matrix A.

- Decomposition: $A = VDV^{T}$, where *D* is a real nonnegative diagonal matrix, and *V* is unitary. V^{T} denotes the matrix transpose of *V*.
- Comment: The diagonal elements of *D* are the nonnegative square roots of the eigenvalues of AA^{H}.
- Comment: *V* may be complex even if *A* is real.
- Comment: This is not a special case of the eigendecomposition.

Singular Value Decomposition

- Applicable to: *m*-by-*n* matrix *A*.
- Decomposition: $A = UDV^{*}$, where *D* is a nonnegative diagonal matrix, and *U* and *V* are unitary matrices, and V^{*} is the conjugate transpose of *V* (or simply the transpose, if *V* contains real numbers only).
- Comment: The diagonal elements of *D* are called the singular values of *A*.
- Comment: Like the eigendecomposition above, the singular value decomposition involves finding basis directions along which matrix multiplication is equivalent to scalar multiplication, but it has greater generality since the matrix under consideration need not be square.
- Uniqueness: the singular values of A are always uniquely determined. U and V need not to be unique in general.

Other Decompositions

Polar Decomposition

- Applicable to: any square complex matrix *A*.
- Decomposition: $A = UP$ (right polar decomposition) or $A = P'U$ (left polar decomposition), where *U* is a unitary matrix and *P* and *P'* are positive semidefinite Hermitian matrices.
- Uniqueness: P is always unique and equal to $\sqrt{A^{*}A}$ (which is always hermitian and positive semidefinite). If A is invertible, then U is unique.
- Comment: Since any Hermitian matrix admits a spectral decomposition with a unitary matrix, P can be written as $P = VDV^{*}$. Since P is positive semidefinite, all elements in D are non-negative. Since the product of two unitary matrices is unitary, taking $W = UV$ one can write $A = U(VDV^{*}) = WDV^{*}$ which is the singular value decomposition. Hence, the existence of the polar decomposition is equivalent to the existence of the singular value decomposition.

Algebraic Polar Decomposition

- Applicable to: square, complex, non-singular matrix *A*.
- Decomposition: $A = QS$, where *Q* is a complex orthogonal matrix and *S* is complex symmetric matrix.
- Uniqueness: If $A^T A$ has no negative real eigenvalues, then the decomposition is unique.
- Comment: The existence of this decomposition is equivalent to AA^T being similar to $A^T A$.
- Comment: A variant of this decomposition is $A = RC$, where *R* is a real matrix and *C* is a circular matrix.

Mostow's Decomposition

- Applicable to: square, complex, non-singular matrix *A*.
- Decomposition: $A = Ue^{iM}e^{S}$, where *U* is unitary, *M* is real anti-symmetric and *S* is real symmetric.
- Comment: The matrix *A* can also be decomposed as $A = U_2 e^{S_2} e^{iM_2}$, where U_2 is unitary, M_2 is real anti-symmetric and S_2 is real symmetric.

Sinkhorn Normal Form

- Applicable to: square real matrix *A* with strictly positive elements.
- Decomposition: $A = D_1 S D_2$, where *S* is doubly stochastic and D_1 and D_2 are real diagonal matrices with strictly positive elements.

Sectoral Decomposition

- Applicable to: square, complex matrix *A* with numerical range contained in the sector $S_\alpha = \{re^{i\theta} \in \mathbb{C} \mid r > 0, |\theta| \le \alpha < \frac{\pi}{2}\}$.
- Decomposition: $A = CZC^H$, where *C* is an invertible complex matrix and $Z = \operatorname{diag}(e^{i\theta_1}, \ldots, e^{i\theta_n})$ with all $|\theta_j| \le \alpha$.

Williamson's Normal Form

- Applicable to: square, positive-definite real matrix *A* with order 2*n*-by-2*n*.
- Decomposition: $A = S^T \operatorname{diag}(D, D) S$, where *S* is a symplectic matrix and *D* is a nonnegative diagonal matrix.

Generalizations

There exist analogues of the SVD, QR, LU and Cholesky factorizations for quasimatrices and cmatrices or continuous matrices. A 'quasimatrix' is, like a matrix, a rectangular scheme whose elements are indexed, but one discrete index is replaced by a continuous index. Likewise, a 'cmatrix', is continuous in both indices. As an example of a cmatrix, one can think of the kernel of an integral operator.

These factorizations are based on early work by Fredholm (1903), Hilbert (1904) and Schmidt (1907).

Generalized Inverse

In mathematics, a generalized inverse of a matrix *A* is a matrix that has some properties of the inverse matrix of *A* but not necessarily all of them. Formally, given a matrix $A \in \mathbb{R}^{n \times m}$ and a matrix $A^g \in \mathbb{R}^{m \times n}$, A^g is a generalized inverse of A if it satisfies the condition $AA^gA = A$.

The purpose of constructing a generalized inverse is to obtain a matrix that can serve as the inverse in some sense for a wider class of matrices than invertible matrices. A generalized inverse exists for an arbitrary matrix, and when a matrix has an inverse, then this inverse is its unique generalized inverse. Some generalized inverses can be defined in any mathematical structure that involves associative multiplication, that is, in a semigroup.

Motivation for the Generalized Inverse

Consider the linear system

$$Ax = y$$

where A is an $n \times m$ matrix and $y \in \mathcal{R}(A)$, the column space of A. If the matrix A is nonsingular, then $x = A^{-1}y$ will be the solution of the system. Note that, if a matrix A is nonsingular, then

$$AA^{-1}A = A.$$

Suppose the matrix A is singular, or $n \neq m$. Then we need a right candidate G of order $m \times n$ such that for all $y \in \mathcal{R}(A)$,

$$AGy = y.$$

That is, Gy is a solution of the linear system $Ax = y$. Equivalently, we need a matrix G of order $m \times n$ such that

$$AGA = A.$$

Hence we can define the generalized inverse as follows: Given $n \times m$ matrix A, an $m \times n$ matrix G is said to be a generalized inverse of A if $AGA = A$.

Construction of Generalized Inverse

The following characterizations are easy to verify:

1. If $A = BC$ is a rank factorization, then $G = C_r^- B_l^-$ is a g-inverse of A, where C_r^- is a right inverse of C and B_l^- is left inverse of B.

2. If $A = P\begin{bmatrix} I_r & 0 \\ 0 & 0 \end{bmatrix}Q$ for any non-singular matrices P and Q, then

   $G = Q^{-1}\begin{bmatrix} I_r & U \\ W & V \end{bmatrix}P^{-1}$ is a generalized inverse of A for arbitrary U, V and W.

3. Let A be of rank r. Without loss of generality, let

$$A = \begin{bmatrix} B & C \\ D & E \end{bmatrix},$$

where $B_{r \times r}$ is the non-singular submatrix of A. Then,

$G = \begin{bmatrix} B^{-1} & 0 \\ 0 & 0 \end{bmatrix}$ is a g-inverse of A.

Types of Generalized Inverses

The Penrose conditions are used to define different generalized inverses for $A \in \mathbb{R}^{n \times m}$ and $A^g \in \mathbb{R}^{m \times n}$:

1. $AA^g A = A$
2. $A^g AA^g = A^g$
3. $(AA^g)^T = AA^g$
4. $(A^g A)^T = A^g A$

If A^g satisfies the first condition, then it is a generalized inverse of A. If it satisfies the first two conditions, then it is a generalized reflexive inverse of A. If it satisfies all four conditions, then it is a Moore–Penrose pseudoinverse of A.

Other kinds of generalized inverse include:

- One-sided inverse (left inverse or right inverse): If the matrix A has dimensions $n \times m$ and is full rank, then use the left inverse if $n > m$ and the right inverse if $n < m$.
    - Left inverse is given by $A_{left}^{-1} = \left(A^T A\right)^{-1} A^T$, i.e., $A_{left}^{-1} A = I_m$, where I_m is the $m \times m$ identity matrix.
    - Right inverse is given by $A_{right}^{-1} = A^T \left(AA^T\right)^{-1}$, i.e., $AA_{right}^{-1} = I_n$, where I_n is the $n \times n$ identity matrix.
- Drazin inverse
- Bott–Duffin inverse
- Moore–Penrose pseudoinverse

Uses

Any generalized inverse can be used to determine whether a system of linear equations has any solutions, and if so to give all of them. If any solutions exist for the $n \times m$ linear system

$$Ax = b,$$

with vector x of unknowns and vector b of constants, all solutions are given by

$$x = A^g b + [I - A^g A]w,$$

parametric on the arbitrary vector w, where A^g is any generalized inverse of A. Solutions exist if and only if $A^g b$ is a solution, that is, if and only if $AA^g b = b$.

References

- Choudhury, Dipa; Horn, Roger A. (April 1987). "A Complex Orthogonal-Symmetric Analog of the Polar Decomposition". SIAM Journal on Algebraic Discrete Methods. 8 (2). doi:10.1137/0608019
- Arnold, Vladimir I.; Cooke, Roger (1992), Ordinary differential equations, Berlin, DE; New York, NY: Springer-Verlag, ISBN 978-3-540-54813-3
- Coppersmith, D.; Winograd, S. (1990). "Matrix multiplication via arithmetic progressions". J. Symbolic Comput. 9: 251–280. doi:10.1016/s0747-7171(08)80013-2
- Baker, Andrew J. (2003), Matrix Groups: An Introduction to Lie Group Theory, Berlin, DE; New York, NY: Springer-Verlag, ISBN 978-1-85233-470-3
- Stewart, G. W. (2011), Fredholm, Hilbert, Schmidt: three fundamental papers on integral equations (PDF), retrieved 2015-01-06

- Ran Raz. On the complexity of matrix product. In Proceedings of the thirty-fourth annual ACM symposium on Theory of computing. ACM Press, 2002. doi:10.1145/509907.509932
- Gilbarg, David; Trudinger, Neil S. (2001), Elliptic partial differential equations of second order (2nd ed.), Berlin, DE; New York, NY: Springer-Verlag, ISBN 978-3-540-41160-4
- Styan, George P. H. (1973), "Hadamard Products and Multivariate Statistical Analysis", Linear Algebra and its Applications, 6: 217–240, doi:10.1016/0024-3795(73)90023-2
- Lang, Serge (2002), Algebra, Graduate Texts in Mathematics, 211 (Revised third ed.), New York: Springer-Verlag, ISBN 978-0-387-95385-4, MR 1878556
- Horn, Roger A.; Merino, Dennis I. (January 1995). "Contragredient equivalence: A canonical form and some applications". Linear Algebra and its Applications. 214. doi:10.1016/0024-3795(93)00056-6
- Stinson, Douglas R. (2005), Cryptography, Discrete Mathematics and its Applications, Chapman & Hall/CRC, ISBN 978-1-58488-508-5
- Townsend, A.; Trefethen, L. N. (2015), "Continuous analogues of matrix factorizations", Proc. R. Soc. A, 471 (2173), doi:10.1098/rspa.2014.0585
- Press, William H.; Flannery, Brian P.; Teukolsky, Saul A.; Vetterling, William T. (2007), Numerical Recipes: The Art of Scientific Computing (3rd ed.), Cambridge University Press, ISBN 978-0-521-88068-8

4

An Integrated Study of Linear Transformations

A linear transformation between two vector spaces $V \rightarrow W$ can be such that the operations of addition and scalar multiplication are preserved. The expected value of a random variable, the identity map on any module are examples of linear values. Linear algebra is best understood in confluence with the major topics listed in the following chapter.

Linear Map

In mathematics, a linear map (also called a linear mapping, linear transformation or, in some contexts, linear function) is a mapping $V \rightarrow W$ between two modules (including vector spaces) that preserves (in the sense defined below) the operations of addition and scalar multiplication.

An important special case is when V = W, in which case the map is called a linear operator, or an endomorphism of V. Sometimes the term linear function has the same meaning as lin*ear map*, while in analytic geometry it does not.

A linear map always maps linear subspaces onto linear subspaces (possibly of a lower dimension); for instance it maps a plane through the origin to a plane, straight line or point. Linear maps can often be represented as matrices, and simple examples include rotation and reflection linear transformations.

In the language of abstract algebra, a linear map is a module homomorphism. In the language of category theory it is a morphism in the category of modules over a given ring.

Definition and First Consequences

Let V and W be vector spaces over the same field K. A function $f: V \rightarrow W$ is said to be a *linear map* if for any two vectors **x** and **y** in V and any scalar α in K, the following two conditions are satisfied:

$f(x+y) = f(x) + f(y)$	additivity
$f(\alpha x) = \alpha f(x)$	homogeneity of degree 1

This is equivalent to requiring the same for any linear combination of vectors, i.e. that for any vectors $x_1, ..., x_m \in V$ and scalars $a_1, ..., a_m \in K$, the following equality holds:

$$f(c_1u_1 + \cdots + c_nu_n) = c_1f(u_1) + \cdots + c_nf(u_n).$$

Denoting the zero elements of the vector spaces V and W by 0_V and 0_W respectively, it follows that $f(0_V) = 0_W$ because letting $a = 0$ in the equation for homogeneity of degree 1,

$$f(0_V) = f(0 \cdot 0_V) = 0 \cdot f(0_V) = 0_W.$$

Occasionally, V and W can be considered to be vector spaces over different fields. It is then necessary to specify which of these ground fields is being used in the definition of "linear". If V and W are considered as spaces over the field K as above, we talk about K-linear maps. For example, the conjugation of complex numbers is an R-linear map **C** → C, but it is not C-linear.

A linear map from V to K (with K viewed as a vector space over itself) is called a linear functional.

These statements generalize to any left-module ${}_RM$ over a ring R without modification, and to any right-module upon reversing of the scalar multiplication.

Examples

- The zero map between two left-modules (or two right-modules) over the same ring is always linear.
- The identity map on any module is a linear operator.
- Any homothecy centered in the origin of a vector space, $v \mapsto cv$ where c is a scalar, is a linear operator. This does not hold in general for modules, where such a map might only be semilinear.
- For real numbers, the map $x \mapsto x^2$ is not linear.
- For real numbers, the map $x \mapsto x + 1$ is not linear (but is an affine transformation; $y = x + 1$ is a linear equation, as the term is used in analytic geometry.)
- If A is a real $m \times n$ matrix, then A defines a linear map from R^n to R^m by sending the column vector $x \in R^n$ to the column vector $Ax \in R^m$. Conversely, any linear map between finite-dimensional vector spaces can be represented in this manner.
- Differentiation defines a linear map from the space of all differentiable functions to the space of all functions. It also defines a linear operator on the space of all smooth functions.
- The (definite) integral over some interval I is a linear map from the space of all real-valued integrable functions on I to R

- The (indefinite) integral (or antiderivative) with a fixed starting point defines a linear map from the space of all real-valued integrable functions on R to the space of all real-valued, differentiable, functions on R. Without a fixed starting point, an exercise in group theory will show that the antiderivative maps to the quotient space of the differentiables over the equivalence relation, "differ by a constant", which yields an identity class of the constant valued functions $\left(\int: I(\mathfrak{R}) \rightarrow D(\mathfrak{R}) / \mathfrak{R}\right)$.
- If V and W are finite-dimensional vector spaces over a field F, then functions that send linear maps $f: V \rightarrow W$ to $\dim_F(W) \times \dim_F(V)$ matrices in the way described in the sequel are themselves linear maps (indeed linear isomorphisms).
- The expected value of a random variable (which is in fact a function, and as such a member of a vector space) is linear, as for random variables X and Y we have $E[X + Y] = E[X] + E[Y]$ and $E[aX] = aE[X]$, but the variance of a random variable is not linear.

Matrices

If V and W are finite-dimensional vector spaces and a basis is defined for each vector space, then every linear map from V to W can be represented by a matrix. This is useful because it allows concrete calculations. Matrices yield examples of linear maps: if A is a real $m \times n$ matrix, then $f(\mathbf{x}) = A\mathbf{x}$ describes a linear map $R^n \rightarrow R^m$.

Let $\{v_1, ..., v_n\}$ be a basis for V. Then every vector $\mathbf{v}$ in V is uniquely determined by the coefficients $c_1, ..., c_n$ in the field R:

$$c_1 v_1 + \cdots + c_n v_n.$$

If $f: V \rightarrow W$ is a linear map,

$$f(c_1 v_1 + \cdots + c_n v_n) = c_1 f(v_1) + \cdots + c_n f(v_n),$$

which implies that the function f is entirely determined by the vectors $f(\mathbf{v}_1), ..., f(\mathbf{v}_n)$. Now let $\{w_1, ..., w_m\}$ be a basis for W. Then we can represent each vector $f(\mathbf{v}_j)$ as

$$f(v_j) = a_{1j} w_1 + \cdots + a_{mj} w_m.$$

Thus, the function f is entirely determined by the values of a_{ij}. If we put these values into an $m \times n$ matrix M, then we can conveniently use it to compute the vector output of f for any vector in V. To get M, every column j of M is a vector

$$(a_{1j}, ..., a_{mj})^T$$

corresponding to $f(v_j)$ as defined above. To define it more clearly, for some column j that corresponds to the mapping $f(v_j)$,

$$M = \begin{pmatrix} & a_{1j} & \\ & \cdot & \\ * & \cdot & * \\ & \cdot & \\ & a_{mj} & \end{pmatrix}$$

where M is the matrix of f. The symbol $*$ denotes that there are other columns which together with column j make up a total of n columns of M. In other words, every column $j = 1, \ldots, n$ has a corresponding vector $f(\mathbf{v}_j)$ whose coordinates $a_{1j}, \ldots, a_{mj}$ are the elements of column j. A single linear map may be represented by many matrices. This is because the values of the elements of a matrix depend on the bases chosen.

The matrices of a linear transformation can be represented visually:

1. Matrix for T relative to B : A
2. Matrix for T relative to B′ : A′
3. Transition matrix from B′ to B : P
4. Transition matrix from B to B′ : P^{-1}

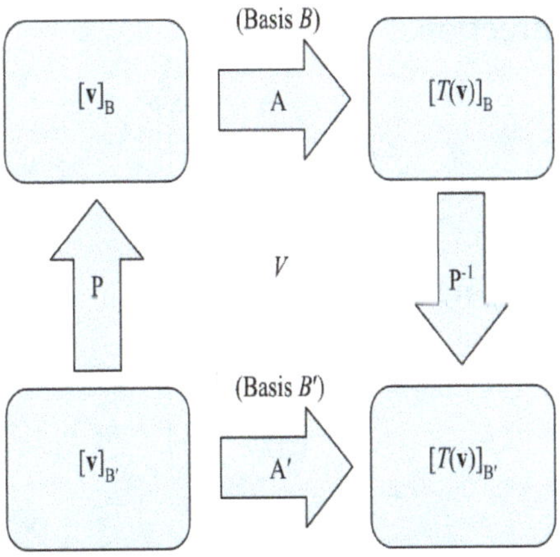

Such that starting in the bottom left corner $[\vec{v}]_{B'}$ and looking for the bottom right corner $[T(\vec{v})]_{B'}$, one would left-multiply—that is, $A'[\vec{v}]_{B'} = [T(\vec{v})]_{B'}$. The equivalent method would be the "longer" method going clockwise from the same point such that $[\vec{v}]_{B'}$ is left-multiplied with $P^{-1}AP$, or $P^{-1}AP[\vec{v}]_{B'} = [T(\vec{v})]_{B'}$.

Examples of Linear Transformation Matrices

In two-dimensional space R^2 linear maps are described by 2×2 real matrices. These are some examples:

- rotation by 90 degrees counterclockwise:

$$A = \begin{pmatrix} 0 & -1 \\ 1 & 0 \end{pmatrix}$$

- rotation by angle θ counterclockwise:

$$A = \begin{pmatrix} \cos\theta & -\sin\theta \\ \sin\theta & \cos\theta \end{pmatrix}$$

- reflection against the x axis:

$$A = \begin{pmatrix} 1 & 0 \\ 0 & -1 \end{pmatrix}$$

- reflection against the y axis:

$$A = \begin{pmatrix} -1 & 0 \\ 0 & 1 \end{pmatrix}$$

- scaling by 2 in all directions:

$$A = \begin{pmatrix} 2 & 0 \\ 0 & 2 \end{pmatrix}$$

- horizontal shear mapping

$$A = \begin{pmatrix} 1 & m \\ 0 & 1 \end{pmatrix}$$

- squeeze mapping:

$$A = \begin{pmatrix} k & 0 \\ 0 & 1/k \end{pmatrix}$$

- projection onto the y axis:

$$A = \begin{pmatrix} 0 & 0 \\ 0 & 1 \end{pmatrix}.$$

Forming New Linear Maps from Given Ones

The composition of linear maps is linear: if $f: V \to W$ and $g: W \to Z$ are linear, then so is their composition $g \circ f: V \to Z$. It follows from this that the class of all vector spaces over a given field K, together with K-linear maps as morphisms, forms a category.

The inverse of a linear map, when defined, is again a linear map.

If $f_1 : V \to W$ and $f_2 : V \to W$ are linear, then so is their pointwise sum $f_1 + f_2$ (which is defined by $(f_1 + f_2)(\mathbf{x}) = f_1(\mathbf{x}) + f_2(\mathbf{x})$).

If $f: V \to W$ is linear and a is an element of the ground field K, then the map af, defined by $(af)(\mathbf{x}) = a(f(\mathbf{x}))$, is also linear.

Thus the set $L(V, W)$ of linear maps from V to W itself forms a vector space over K, sometimes denoted Hom(V, W). Furthermore, in the case that $V = W$, this vector space (denoted End(V)) is an associative algebra under composition of maps, since the composition of two linear maps is again a linear map, and the composition of maps is always associative. This case is discussed in more detail below.

Given again the finite-dimensional case, if bases have been chosen, then the composition of linear maps corresponds to the matrix multiplication, the addition of linear maps corresponds to the matrix addition, and the multiplication of linear maps with scalars corresponds to the multiplication of matrices with scalars.

Endomorphisms and Automorphisms

A linear transformation $f: V \to V$ is an endomorphism of V; the set of all such endomorphisms End(V) together with addition, composition and scalar multiplication as defined above forms an associative algebra with identity element over the field K (and in particular a ring). The multiplicative identity element of this algebra is the identity map id: $V \to V$.

An endomorphism of V that is also an isomorphism is called an automorphism of V. The composition of two automorphisms is again an automorphism, and the set of all automorphisms of V forms a group, the automorphism group of V which is denoted by Aut(V) or GL(V). Since the automorphisms are precisely those endomorphisms which possess inverses under composition, Aut(V) is the group of units in the ring End(V).

If V has finite dimension n, then End(V) is isomorphic to the associative algebra of all $n \times n$ matrices with entries in K. The automorphism group of V is isomorphic to the general linear group GL(n, K) of all $n \times n$ invertible matrices with entries in K.

Kernel, Image and the Rank–nullity Theorem

If $f: V \to W$ is linear, we define the kernel and the image or range of f by

$$\ker(f) = \{x \in V : f(x) = 0\}$$
$$\operatorname{im}(f) = \{w \in W : w = f(x), x \in V\}$$

ker(*f*) is a subspace of *V* and im(*f*) is a subspace of *W*. The following dimension formula is known as the rank–nullity theorem:

$$\dim(\ker(f)) + \dim(\operatorname{im}(f)) = \dim(V).$$

The number dim(im(*f*)) is also called the *rank of f* and written as rank(*f*), or sometimes, ρ(*f*); the number dim(ker(*f*)) is called the *nullity of f* and written as null(*f*) or ν(*f*). If *V* and *W* are finite-dimensional, bases have been chosen and *f* is represented by the matrix *A*, then the rank and nullity of *f* are equal to the rank and nullity of the matrix *A*, respectively.

Cokernel

A subtler invariant of a linear transformation $f : V \to W$ is the *co*kernel, which is defined as

$$\operatorname{coker} f := W / f(V) = W / \operatorname{im}(f).$$

This is the *dual* notion to the kernel: just as the kernel is a *sub*space of the *domain*, the co-kernel is a *quotient* space of the *target*. Formally, one has the exact sequence

$$0 \to \ker f \to V \to W \to \operatorname{coker} f \to 0.$$

These can be interpreted thus: given a linear equation *f*(**v**) = **w** to solve,

- the kernel is the space of *solutions* to the *homogeneous* equation *f*(**v**) = 0, and its dimension is the number of *degrees of freedom* in a solution, if it exists;
- the co-kernel is the space of *constraints* that must be satisfied if the equation is to have a solution, and its dimension is the number of constraints that must be satisfied for the equation to have a solution.

The dimension of the co-kernel and the dimension of the image (the rank) add up to the dimension of the target space. For finite dimensions, this means that the dimension of the quotient space *W*/*f*(*V*) is the dimension of the target space minus the dimension of the image.

As a simple example, consider the map *f*: $R^2 \to R^2$, given by *f*(*x*, *y*) = (0, *y*). Then for an equation *f*(*x*, *y*) = (*a*, *b*) to have a solution, we must have *a* = 0 (one constraint), and in that case the solution space is (*x*, *b*) or equivalently stated, (0, *b*) + (*x*, 0), (one degree of freedom). The kernel may be expressed as the subspace (*x*, 0) < *V*: the value of *x* is the freedom in a solution – while the cokernel may be expressed via the map $W \to R$, $(a, b) \mapsto (a)$: given a vector (*a*, *b*), the value of *a* is the *obstruction* to there being a solution.

An example illustrating the infinite-dimensional case is afforded by the map f: $\mathrm{R}^\infty \to \mathrm{R}^\infty$, $\{a_n\} \mapsto \{b_n\}$ with $b_1 = 0$ and $b_{n+1} = a_n$ for $n > 0$. Its image consists of all sequences with first element 0, and thus its cokernel consists of the classes of sequences with identical first element. Thus, whereas its kernel has dimension 0 (it maps only the zero sequence to the zero sequence), its co-kernel has dimension 1. Since the domain and the target space are the same, the rank and the dimension of the kernel add up to the same sum as the rank and the dimension of the co-kernel ($\aleph_0 + 0 = \aleph_0 + 1$), but in the infinite-dimensional case it cannot be inferred that the kernel and the co-kernel of an endomorphism have the same dimension ($0 \neq 1$). The reverse situation obtains for the map h: $\mathrm{R}^\infty \to \mathrm{R}^\infty$, $\{a_n\} \mapsto \{c_n\}$ with $c_n = a_{n+1}$. Its image is the entire target space, and hence its co-kernel has dimension 0, but since it maps all sequences in which only the first element is non-zero to the zero sequence, its kernel has dimension 1.

Index

For a linear operator with finite-dimensional kernel and co-kernel, one may define *index* as:

$$ind\, f := \dim \ker f - \dim coker\, f,$$

namely the degrees of freedom minus the number of constraints.

For a transformation between finite-dimensional vector spaces, this is just the difference $\dim(V) - \dim(W)$, by rank–nullity. This gives an indication of how many solutions or how many constraints one has: if mapping from a larger space to a smaller one, the map may be onto, and thus will have degrees of freedom even without constraints. Conversely, if mapping from a smaller space to a larger one, the map cannot be onto, and thus one will have constraints even without degrees of freedom.

The index of an operator is precisely the Euler characteristic of the 2-term complex $0 \to V \to W \to 0$. In operator theory, the index of Fredholm operators is an object of study, with a major result being the Atiyah–Singer index theorem.

Algebraic Classifications of Linear Transformations

No classification of linear maps could hope to be exhaustive. The following incomplete list enumerates some important classifications that do not require any additional structure on the vector space.

Let V and W denote vector spaces over a field, F. Let T: $V \to W$ be a linear map.

- T is said to be *injective* or a *monomorphism* if any of the following equivalent conditions are true:
    - T is one-to-one as a map of sets.

- $\ker T = \{0_V\}$
- T is monic or left-cancellable, which is to say, for any vector space U and any pair of linear maps $R: U \to V$ and $S: U \to V$, the equation $TR = TS$ implies $R = S$.
- T is left-invertible, which is to say there exists a linear map $S: W \to V$ such that ST is the identity map on V.

- T is said to be *surjective* or an *epimorphism* if any of the following equivalent conditions are true:
  - T is onto as a map of sets.
  - $\operatorname{coker} T = \{0_W\}$
  - T is epic or right-cancellable, which is to say, for any vector space U and any pair of linear maps $R: W \to U$ and $S: W \to U$, the equation $RT = ST$ implies $R = S$.
  - T is right-invertible, which is to say there exists a linear map $S: W \to V$ such that TS is the identity map on W.
- T is said to be an *isomorphism* if it is both left- and right-invertible. This is equivalent to T being both one-to-one and onto (a bijection of sets) or also to T being both epic and monic, and so being a bimorphism.
- If $T: V \to V$ is an endomorphism, then:
  - If, for some positive integer n, the n-th iterate of T, T^n, is identically zero, then T is said to be nilpotent.
  - If $T^2 = T$, then T is said to be idempotent
  - If $T = kI$, where k is some scalar, then T is said to be a scaling transformation or scalar multiplication map.

Change of Basis

Given a linear map which is an endomorphism whose matrix is A, in the basis B of the space it transforms vector coordinates [u] as [v] = A[u]. As vectors change with the inverse of B (vectors are contravariant) its inverse transformation is [v] = B[v'].

Substituting this in the first expression

$$B[v'] = AB[u']$$

hence

$$[v'] = B^{-1}AB[u'] = A'[u'].$$

Therefore, the matrix in the new basis is $A' = B^{-1}AB$, being B the matrix of the given basis.

Therefore, linear maps are said to be 1-co 1-contra -variant objects, or type (1, 1) tensors.

Continuity

A *linear transformation* between topological vector spaces, for example normed spaces, may be continuous. If its domain and codomain are the same, it will then be a continuous linear operator. A linear operator on a normed linear space is continuous if and only if it is bounded, for example, when the domain is finite-dimensional. An infinite-dimensional domain may have discontinuous linear operators.

An example of an unbounded, hence discontinuous, linear transformation is differentiation on the space of smooth functions equipped with the supremum norm (a function with small values can have a derivative with large values, while the derivative of 0 is 0). For a specific example, $\sin(nx)/n$ converges to 0, but its derivative $\cos(nx)$ does not, so differentiation is not continuous at 0 (and by a variation of this argument, it is not continuous anywhere).

Applications

A specific application of linear maps is for geometric transformations, such as those performed in computer graphics, where the translation, rotation and scaling of 2D or 3D objects is performed by the use of a transformation matrix. Linear mappings also are used as a mechanism for describing change: for example in calculus correspond to derivatives; or in relativity, used as a device to keep track of the local transformations of reference frames.

Another application of these transformations is in compiler optimizations of nested-loop code, and in parallelizing compiler techniques.

Eigenvalues and Eigenvectors

Given a linear operator $X \xrightarrow{A} X$, a scalar $\lambda \in \mathbb{F}$ is called an eigen value of A iff the equation $Ax = \lambda x$ has a non-zero solutions; put another way, $\ker(A - \lambda I) := \{x \in \mathbb{X} \mid Ax = \lambda x\}$ is a non-zero subspace of X. Thus for $\dim X = n < \infty$, λ is an eigen space of A iff λ is a solution of the equation.

$$x_A(\lambda) := \det(A - \lambda I) = 0 \qquad (1)$$

(Since an operator $X \xrightarrow{B} X$ is invertible iff B is an isomorphism which happens, iff for some matrix representation b for B, we have b invertible, i.e. $\det b \neq 0$).

The equation (1) is

$$x_A(\lambda) := a_0 + a_1\lambda + \ldots + a_n\lambda^n = 0$$

which is obtained from the polynomial

$$x_A(\theta) := a_0 + a_1\theta + \ldots + a_n\theta^n \in \mathbb{F}\theta$$

for some coefficients $a_0, \ldots, a_n \in$. This polynomial $x_A(\theta)$ is known as the characteristic polynomial of the operator A. When a is some matrix representation of A, there is the same polynomial $x_a(\theta)$ though we call it the characteristic polynomial of the matrix a. To see this, suppose b is some other matrix representation of A. Then $b = p^{-1}ap$ for some invertible matrix (∵ a change-of-basis matrix is the same thing as an invertible matrix.) Then

$$\begin{aligned} x_b(\lambda) &= \det(b - \lambda I_n) \\ &= \det(p^{-1}ap - \lambda I_n) \\ &= \det(p^{-1}(a - \lambda I_n)p) \\ &= (\det(p^{-1}))\det(a - \lambda I_n)(\det(p)) \\ &= (\det p)^{-1} x_a(\lambda)(\det p) \\ &= x_a\lambda \end{aligned}$$

Thus λ is an eigenvalue of A

iff λ is an eigenvalue of a

iff $x_a(\lambda) = 0$

iff $x_b(\lambda) = 0$

iff λ is an eigenvalue of b

iff λ is an eigenvalue of B.

We note that the polynomial $x_A(\theta) := a_0 + a_1\theta + \ldots + a_n\theta^n$ is in fact a monic polynomial i.e. $a_n = 1$. Supposing $\mathbb{F} = \mathbb{C}$, this can be written $x_A(\theta) := (\theta - \lambda_1)^{n_1} \ldots (\theta - \lambda_k)^{n_k}$ where $\lambda_1, \ldots, \lambda_k$ exhausts all the eigen values of A; further, $n_1 + \ldots + n_k = n = \dim X$.

Comment The definition supplied in (1) is obviously concealing something: What ia after all the meaning of the determinant of an operator $T \in L(X, X)$(in this case $T = A - \lambda I$)?

One may then say, in view of the fact (established above) that since $\det(a - \lambda I_n) = \det(b - \lambda I_n)$ for any two matrices a, b representing the operator $A \in L(X, X)$, the characteristic polynomial $x_A(\theta)$ of A is the same as $x_a(\theta) = \det(a - \lambda I_n)$ for any matrix a representing A. This again is problematic since it is far from clear what the meaning of the determinant of a square matrix is. However, since you have been working with determinants anyway, this may be taken as a working tool, made available off the shelf.

A clean treatment of these things is out of question here but once you have suppressed your mathematical conscience, there is nothing easier than this topic in linear algebra; as you know it from your school days.

The collection $\{\lambda_1, \ldots, \lambda_k\} \subseteq F$ mentioned in 1 for the linear transformation $X \xrightarrow{A} X$, $\dim X = n < \infty$, is called the eigen spectrum of A; we shall denote it by A_e; its members are, as mentioned before, are called eigen values of A. Since there are no 'spectral values' other than these eigenvalues for the case $\dim X = n < \infty$, a course on finite dimensional vector spaces may, and usually does, call it simply the 'spectrum' of A. But using the correct name never hurts. The subspace $\ker(A - \lambda I)$ of X is called the eigen space of X corresponding to the eigenvalue λ.

The spectrum A_e may be empty: for instance if $\mathbb{F} = \mathbb{R}$, $X = \mathbb{R}^2$, $\left\{ e_1 = \begin{bmatrix} 1 \\ 0 \end{bmatrix}, e_2 = \begin{bmatrix} 0 \\ 1 \end{bmatrix} \right\}$ is chosen as a basis, $X \xrightarrow{A} X$ is defined by $A(e_1) = -e_2, A(e_2) = e_1$ then $x_A(\theta) = x_a(\theta) = \theta^2 + 1$. (Verify that the matrix a with respect to the basis will be $\begin{bmatrix} 0 & 1 \\ -1 & 0 \end{bmatrix}$ since its columns are to be given by $A(e_1) = -e_2, A(e_2) = e_1$ then

$$x_A(\lambda) = \begin{bmatrix} -\lambda & 1 \\ -1 & -\lambda \end{bmatrix} = -(\lambda^2 + 1) \text{ and } x_A(\theta) = 0 \text{ has no roots in } \mathbb{R}.$$

But this will not be the case if the field is algebraically closed; for instance if $\mathbb{F} = \mathbb{C}$. Our illustrations be almost always for the case $\mathbb{F} = \mathbb{C}$.

The minimal polynomial of a linear operator $X \xrightarrow{A} X$, $\dim X = n < \infty$, is defined uniquely by the requirements (denoting this by $\mu_A(\theta)$).

1. $\mu_A(A) = 0 \in L(X, X)$
2. $q(A) = 0, q \neq 0 \Rightarrow \deg q \geq \deg \mu_A(\theta)$ for any $q = q(\theta) \in \mathbb{F}[\theta]$, and
3. $\mu_A(\theta)$ is a monic polynomial

The existence and uniqueness of $\mu_A(\theta)$ can be seen as follows: We consider the collection

$$\mathfrak{I} := \left\{ \mathbb{F}[\theta] \middle| X \xrightarrow{f(A)} X = 0 \in L(X, X) \right\}$$

If f (θ), g(θ) $\in \Im$ then clearly α f(θ) + β g(θ) and f (θ)g(θ) $\in \Im$ for any given α, β ∈ $\mathbb{F}$. This means that $\Im$ is an ideal in the ring of polynomials; say it is the annihilator of A and denote it by Annh(A). Since dim L(X, X) = n² < ∞, the n² + 1 operators Id, A,..., A^{n^2} cannot be linearly independent in the vector space L(X, X) and there must exist scalars $a_o, \ldots, a_n$ such that, with A^o = Id,

$$a_0A^0 + \ldots + a_{n^2}A^{n^2} = 0,\ a_j \neq 0 \text{ for some } j, 1 \leq j \leq n^2$$

Therefore, Annh(A) contains some non-zero polynomial a(θ), degree a(θ) ≤ n² and thus Annh(A) ≠ {o}.

Now every ideal in $\mathbb{F}$ [θ] is a principal ideal and hence there exists a single generator of Annh(A), there will be exactly one which is monic. This is our $\mu_A(\theta)$.

Thus (3) fixes $\mu_A(\theta)$ uniquely while (1) and (2) determine it up to a scalar multiple.

More information about the minimal polynomial is available from the Cayley-Hamilton Theorem which says that the characteristic polynomial $x_A(\theta)$ ∈ Annh(A). We will accept this result off the shelf; any clean treatment needs more machinery than we can afford to build her. But note that since degree But note that since degree ($x_A(\theta)$) = n, and $\mu_A(\theta)$ must divide $x_A(\theta)$(∴ $\mu_A(\theta)$ divides each f (θ) ∈ Annh(A)), so that deg$\mu_A(\theta)$ ≤ n. Then since

$$x_A(\theta) = (\theta - \lambda_1)^{n_1} \ldots (\theta - \lambda_k)^{n_k}$$

we get

$$\mu_A(\theta) = (\theta - \lambda_1)^{p_1} \ldots (\theta - \lambda_k)^{p_k}, 1 \leq p_i \leq n_i \text{ for } 1 \leq i \leq k.$$

Thus for $X \xrightarrow{A} X, \lambda \in \mathbb{F}$.

We note that an eigenvector of A is a non zero vector (∵ Ax = λx must have a non zero solution by definition.) Assuming that whenever $\lambda_1,\ldots,\lambda_{r-1}$ are distinct eigen values of A, the corresponding eigen vectors $x_1,\ldots,x_{r-1}$ are linearly independent, we find that in

$$\alpha_1 x_1 + \ldots + \alpha_{r-1} x_{r-1} + \alpha_r x_r = 0$$

subject to $\alpha_1,\ldots,\alpha_r$ not all zero, we must have αr ≠ o. (∵ otherwise, $\alpha_1 x_1 + \ldots + \alpha_{r-1} x_{r-1} = 0$ and this forces $\alpha_1 = \ldots = \alpha_{r-1} = 0$ since $x_1,\ldots,x_{r-1}$ are linearly independent by hypothesis; as $\alpha_r = 0$ is already there, the condition "subject to $\alpha_1, \ldots, \alpha_r$ not all zero" is violated). If x_r is an eigenvector with λ_r as eigenvalue (so that $Ax_r = \lambda_r x_r$) with $\lambda_r \neq \lambda_j$ for 1 ≤ j ≤ r – 1, we obtain

$$\begin{aligned}0 &= A0\\ &= A(\alpha_1 x_1 + \ldots + \alpha_r x_r)\\ &= \alpha_1 Ax_1 + \ldots + \alpha_r Ax_r\\ &= \alpha_1\lambda_1 x_1 + \ldots + \alpha_r\lambda_r x_r\end{aligned}$$

Now $\lambda_r = 0$ means no $\lambda_j = 0$ for $1 \le j \le r - 1$ ($\because \lambda_r \ne \lambda_j$ for $1 \le j \le r - 1$) but then we are left with $0 = \beta_1 x_1 + \ldots + \beta_{r-1} x_{r-1}$ where $\beta_j = \alpha_j\lambda_j$ for $1 \le j \le r - 1$, which forces $\beta_j = 0$ for each j which forces $\alpha_j = 0$ for each j(since $\lambda_1,\ldots,\lambda_{r-1}$ are non-zero). But then $\alpha_r = 0$ must be true and "subject to $\alpha_1,\ldots,\alpha_r$ not all zero" is again violated. If $\lambda_r \ne 0$, we have

$-\alpha_r x_r = \frac{1}{\lambda_r}[\alpha_1\lambda_1 x_1 + \ldots + \alpha_{r-1}\lambda_{r-1}x_{r-1}]$ which means $(\lambda_r - \lambda_1)\alpha_1 x_1 + \ldots + (\lambda_r - \lambda_{r-1})\alpha_{r-1}x_{r-1} = 0$.

Since $x_1,\ldots,x_{r-1}$ are assumed linearly independent, we have $(\lambda_r - \lambda_j)\alpha_j = 0$ for $1 \le j \le r - 1$ which forces $\alpha_j = 0$ for $1 \le j \le r - 1$. But if $\alpha_1,\ldots,\alpha_{r-1} = 0$, we get $\alpha_r = 0$ again, which is a contradiction as we know. We conclude that it is not possible to have $0 = \alpha_1 x_1 + \ldots + \alpha_r x_r$ unless $\alpha_1 = \ldots = \alpha_r = 0$ when $x_1,\ldots,x_r$ are eigenvectors corresponding to distinct eigenvalues $\lambda_1,\ldots,\lambda_r$ i.e. the vectors $\{x_1,\ldots,x_{r-1}\}$ are. Since the empty set is always linearly independent (which corresponds to r = 0 in the preceding); we conclude

The Eigenvectors Associated with Distinct Eigenvalues of $X \to AX$ are Linearly Independent

and therefore

If $\lambda_1,\ldots,\lambda_r$ are distinct eigenvalues of $X \to AX$, the sum

$$\oplus_{j=1}^{r} \ker(A - \lambda_r \mathrm{Id})$$

is a direct sum.

Diagonalizable Matrix

In linear algebra, a square matrix A is called diagonalizable if it is similar to a diagonal matrix, i.e., if there exists an invertible matrix P such that $P^{-1}AP$ is a diagonal matrix. If V is a finite-dimensional vector space, then a linear map $T : V \to V$ is called diagonalizable if there exists an ordered basis of V with respect to which T is represented by a diagonal matrix. Diagonalization is the process of finding a corresponding diagonal matrix for a diagonalizable matrix or linear map. A square matrix that is not diagonalizable is called *defective*.

Diagonalizable matrices and maps are of interest because diagonal matrices are especially easy to handle: their eigenvalues and eigenvectors are known, one can raise a diagonal matrix to a power by simply raising the diagonal entries to that same power, and the determinant of a diagonal matrix is simply the product of all entries. Geometrically, a diagonalizable matrix is an inhomogeneous dilation (or *anisotropic scaling*) – it scales the space, as does a *homogeneous dilation*, but by a different factor in each direction, determined by the scale factors on each axis (diagonal entries).

Characterization

The fundamental fact about diagonalizable maps and matrices is expressed by the following:

- An $n \times n$ matrix A over the field F is diagonalizable if and only if the sum of the dimensions of its eigenspaces is equal to n, which is the case if and only if there exists a basis of F^n consisting of eigenvectors of A. If such a basis has been found, one can form the matrix P having these basis vectors as columns, and $P^{-1}AP$ will be a diagonal matrix. The diagonal entries of this matrix are the eigenvalues of A.
- A linear map $T: V \to V$ is diagonalizable if and only if the sum of the dimensions of its eigenspaces is equal to $\dim(V)$, which is the case if and only if there exists a basis of V consisting of eigenvectors of T. With respect to such a basis, T will be represented by a diagonal matrix. The diagonal entries of this matrix are the eigenvalues of T.

Another characterization: A matrix or linear map is diagonalizable over the field F if and only if its minimal polynomial is a product of distinct linear factors over F. (Put in another way, a matrix is diagonalizable if and only if all of its elementary divisors are linear.)

The following sufficient (but not necessary) condition is often useful.

- An $n \times n$ matrix A is diagonalizable over the field F if it has n distinct eigenvalues in F, i.e. if its characteristic polynomial has n distinct roots in F; however, the converse may be false. Let us consider

$$\begin{bmatrix} -1 & 3 & -1 \\ -3 & 5 & -1 \\ -3 & 3 & 1 \end{bmatrix},$$

which has eigenvalues 1, 2, 2 (not all distinct) and is diagonalizable with diagonal form (similar to A)

$$\begin{bmatrix} 1 & 0 & 0 \\ 0 & 2 & 0 \\ 0 & 0 & 2 \end{bmatrix}$$

and change of basis matrix P

$$\begin{bmatrix} 1 & 1 & -1 \\ 1 & 1 & 0 \\ 1 & 0 & 3 \end{bmatrix}.$$

The converse fails when A has an eigenspace of dimension higher than 1. In this example, the eigenspace of A associated with the eigenvalue 2 has dimension 2.

- A linear map $T: V \to V$ with $n = \dim(V)$ is diagonalizable if it has n distinct eigenvalues, i.e. if its characteristic polynomial has n distinct roots in F.

Let A be a matrix over F. If A is diagonalizable, then so is any power of it. Conversely, if A is invertible, F is algebraically closed, and A^n is diagonalizable for some n that is not an integer multiple of the characteristic of F, then A is diagonalizable. Proof: If A^n is diagonalizable, then A is annihilated by some polynomial $(x^n - \lambda_1)\cdots(x^n - \lambda_k)$, which has no multiple root (since $\lambda_j \neq 0$) and is divided by the minimal polynomial of A.

As a rule of thumb, over C almost every matrix is diagonalizable. More precisely: the set of complex $n \times n$ matrices that are *not* diagonalizable over C, considered as a subset of $\mathbf{C}^{n\times n}$, has Lebesgue measure zero. One can also say that the diagonalizable matrices form a dense subset with respect to the Zariski topology: the complement lies inside the set where the discriminant of the characteristic polynomial vanishes, which is a hypersurface. From that follows also density in the usual (*strong*) topology given by a norm. The same is not true over R.

The Jordan–Chevalley decomposition expresses an operator as the sum of its semisimple (i.e., diagonalizable) part and its nilpotent part. Hence, a matrix is diagonalizable if and only if its nilpotent part is zero. Put in another way, a matrix is diagonalizable if each block in its Jordan form has no nilpotent part; i.e., each "block" is a one-by-one matrix.

Diagonalization

If a matrix A can be diagonalized, that is,

$$P^{-1}AP = \begin{pmatrix} \lambda_1 & & & \\ & \lambda_2 & & \\ & & \ddots & \\ & & & \lambda_n \end{pmatrix},$$

then:

$$AP = P\begin{pmatrix} \lambda_1 & & & \\ & \lambda_2 & & \\ & & \ddots & \\ & & & \lambda_n \end{pmatrix}.$$

Writing *P* as a block matrix of its column vectors $\vec{\alpha}_i$

$$P = (\vec{\alpha}_1 \quad \vec{\alpha}_2 \quad \cdots \quad \vec{\alpha}_n),$$

the above equation can be rewritten as

$$A\vec{\alpha}_i = \lambda_i \vec{\alpha}_i \qquad (i = 1, 2, \cdots, n).$$

So the column vectors of *P* are right eigenvectors of *A*, and the corresponding diagonal entry is the corresponding eigenvalue. The invertibility of *P* also suggests that the eigenvectors are linearly independent and form a basis of F^n. This is the necessary and sufficient condition for diagonalizability and the canonical approach of diagonalization. The row vectors of P^{-1} are the left eigenvectors of *A*.

When the matrix *A* is a Hermitian matrix (resp. symmetric matrix), eigenvectors of *A* can be chosen to form an orthonormal basis of C^n (resp. R^n). Under such circumstance *P* will be a unitary matrix (resp. orthogonal matrix) and P^{-1} equals the conjugate transpose (resp. transpose) of *P*.

In practice, matrices are diagonalized numerically using computers. Many algorithms exist to accomplish this.

Simultaneous Diagonalization

A set of matrices is said to be *simultaneously diagonalizable* if there exists a single invertible matrix *P* such that $P^{-1}AP$ is a diagonal matrix for every *A* in the set. The following theorem characterises simultaneously diagonalisable matrices: A set of diagonalizable matrices commutes if and only if the set is simultaneously diagonalisable.

The set of all $n \times n$ diagonalisable matrices (over **C**) with $n > 1$ is not simultaneously diagonalisable. For instance, the matrices

$$\begin{bmatrix} 1 & 0 \\ 0 & 0 \end{bmatrix} \text{ and } \begin{bmatrix} 1 & 1 \\ 0 & 0 \end{bmatrix}$$

are diagonalizable but not simultaneously diagonalizable because they do not commute.

A set consists of commuting normal matrices if and only if it is simultaneously diagonalisable by a unitary matrix; that is, there exists a unitary matrix *U* such that U^*AU is diagonal for every *A* in the set.

In the language of Lie theory, a set of simultaneously diagonalisable matrices generate a toral Lie algebra.

Examples

Diagonalizable Matrices

- Involutions are diagonalizable over the reals (and indeed any field of characteristic not 2), with ±1 on the diagonal.
- Finite order endomorphisms are diagonalisable over **C** (or any algebraically closed field where the characteristic of the field does not divide the order of the endomorphism) with roots of unity on the diagonal. This follows since the minimal polynomial is separable, because the roots of unity are distinct.
- Projections are diagonalizable, with 0s and 1s on the diagonal.
- Real symmetric matrices are diagonalizable by orthogonal matrices; i.e., given a real symmetric matrix A, $Q^T A Q$ is diagonal for some orthogonal matrix Q. More generally, matrices are diagonalizable by unitary matrices if and only if they are normal. In the case of the real symmetric matrix, we see that $A = A^T$, so clearly $AA^T = A^T A$ holds. Examples of normal matrices are real symmetric (or skew-symmetric) matrices (e.g. covariance matrices) and Hermitian matrices (or skew-Hermitian matrices).

Matrices that are not Diagonalizable

In general, a rotation matrix is not diagonalizable over the reals, but all rotation matrices are diagonalizable over the complex field. Even if a matrix is not diagonalizable, it is always possible to "do the best one can", and find a matrix with the same properties consisting of eigenvalues on the leading diagonal, and either ones or zeroes on the superdiagonal - known as Jordan normal form.

Some matrices are not diagonalizable over any field, most notably nonzero nilpotent matrices. This happens more generally if the algebraic and geometric multiplicities of an eigenvalue do not coincide. For instance, consider

$$C = \begin{bmatrix} 0 & 1 \\ 0 & 0 \end{bmatrix}.$$

This matrix is not diagonalizable: there is no matrix U such that $U^{-1}CU$ is a diagonal matrix. Indeed, C has one eigenvalue (namely zero) and this eigenvalue has algebraic multiplicity 2 and geometric multiplicity 1.

Some real matrices are not diagonalizable over the reals. Consider for instance the matrix

$$B = \begin{bmatrix} 0 & 1 \\ -1 & 0 \end{bmatrix}.$$

The matrix *B* does not have any real eigenvalues, so there is no real matrix *Q* such that $Q^{-1}BQ$ is a diagonal matrix. However, we can diagonalize *B* if we allow complex numbers. Indeed, if we take

$$Q = \begin{bmatrix} 1 & i \\ i & 1 \end{bmatrix},$$

then $Q^{-1}BQ$ is diagonal.

Note that the above examples show that the sum of diagonalizable matrices need not be diagonalizable.

How to Diagonalize a Matrix

Consider a matrix

$$A = \begin{bmatrix} 1 & 2 & 0 \\ 0 & 3 & 0 \\ 2 & -4 & 2 \end{bmatrix}.$$

This matrix has eigenvalues

$$\lambda_1 = 3, \quad \lambda_2 = 2, \quad \lambda_3 = 1.$$

A is a 3×3 matrix with 3 different eigenvalues; therefore, it is diagonalizable. Note that if there are exactly *n* distinct eigenvalues in an *n*×*n* matrix then this matrix is diagonalizable.

These eigenvalues are the values that will appear in the diagonalized form of matrix *A*, so by finding the eigenvalues of *A* we have diagonalized it. We could stop here, but it is a good check to use the eigenvectors to diagonalize *A*.

The eigenvectors of *A* are

$$v_1 = \begin{bmatrix} -1 \\ -1 \\ 2 \end{bmatrix}, \quad v_2 = \begin{bmatrix} 0 \\ 0 \\ 1 \end{bmatrix}, \quad v_3 = \begin{bmatrix} -1 \\ 0 \\ 2 \end{bmatrix}.$$

One can easily check that $Av_k = \lambda_k v_k$.

Now, let *P* be the matrix with these eigenvectors as its columns:

$$P = \begin{bmatrix} -1 & 0 & -1 \\ -1 & 0 & 0 \\ 2 & 1 & 2 \end{bmatrix}.$$

Note that there is no preferred order of the eigenvectors in *P*; changing the order of the eigenvectors in *P* just changes the order of the eigenvalues in the diagonalized form of *A*.

Then *P* diagonalizes *A*, as a simple computation confirms, having calculated P^{-1} using any suitable method:

$$P^{-1}AP = \begin{bmatrix} 0 & -1 & 0 \\ 2 & 0 & 1 \\ -1 & 1 & 0 \end{bmatrix} \begin{bmatrix} 1 & 2 & 0 \\ 0 & 3 & 0 \\ 2 & -4 & 2 \end{bmatrix} \begin{bmatrix} -1 & 0 & -1 \\ -1 & 0 & 0 \\ 2 & 1 & 2 \end{bmatrix} = \begin{bmatrix} 3 & 0 & 0 \\ 0 & 2 & 0 \\ 0 & 0 & 1 \end{bmatrix}.$$

In fact, this follows abstractly from the fact that for the standard basis e_1, e_2, e_3 we have

$$P^{-1}APe_k = P^{-1}Av_k = P^{-1}\lambda_k v_k = \lambda_k e_k.$$

where we make use of the fact that $Pe_k = v_k$ is the k-th column of P and hence $P^{-1}v_k = e_k$. Note that the eigenvalues λ_k appear in the diagonal matrix.

An application

Diagonalization can be used to compute the powers of a matrix *A* efficiently, provided the matrix is diagonalizable. Suppose we have found that

$$P^{-1}AP = D \Rightarrow PP^{-1}APP^{-1} = PDP^{-1} \Rightarrow A = PDP^{-1}$$

where D is a diagonal matrix. Then, as the matrix product is associative,

$$\begin{aligned} A^k &= (PDP^{-1})^k = (PDP^{-1})\cdot(PDP^{-1})\cdots(PDP^{-1}) \\ &= PD(P^{-1}P)D(P^{-1}P)\cdots(P^{-1}P)DP^{-1} \\ &= PD^kP^{-1} \end{aligned}$$

and the latter is easy to calculate since it only involves the powers of a diagonal matrix. This approach can be generalized to matrix exponential and other matrix functions since they can be defined as power series.

This is particularly useful in finding closed form expressions for terms of linear recursive sequences, such as the Fibonacci numbers.

Particular Application

For example, consider the following matrix:

$$M = \begin{bmatrix} a & b-a \\ 0 & b \end{bmatrix}.$$

Calculating the various powers of *M* reveals a surprising pattern:

$$M^2 = \begin{bmatrix} a^2 & b^2 - a^2 \\ 0 & b^2 \end{bmatrix}, \quad M^3 = \begin{bmatrix} a^3 & b^3 - a^3 \\ 0 & b^3 \end{bmatrix}, \quad M^4 = \begin{bmatrix} a^4 & b^4 - a^4 \\ 0 & b^4 \end{bmatrix}, \quad \ldots$$

The above phenomenon can be explained by diagonalizing *M*. To accomplish this, we need a basis of R^2 consisting of eigenvectors of *M*. One such eigenvector basis is given by

$$u = \begin{bmatrix} 1 \\ 0 \end{bmatrix} = e_1, \quad v = \begin{bmatrix} 1 \\ 1 \end{bmatrix} = e_1 + e_2,$$

where $\mathbf{e}_i$ denotes the standard basis of R^n. The reverse change of basis is given by

$$e_1 = u, \qquad e_2 = v - u.$$

Straightforward calculations show that

$$Mu = au, \qquad Mv = bv.$$

Thus, *a* and *b* are the eigenvalues corresponding to **u** and **v**, respectively. By linearity of matrix multiplication, we have that

$$M^n u = a^n u, \qquad M^n v = b^n v.$$

Switching back to the standard basis, we have

$$M^n e_1 = M^n u = a^n e_1,$$

$$M^n e_2 = M^n (v - u) = b^n v - a^n u = (b^n - a^n) e_1 + b^n e_2.$$

The preceding relations, expressed in matrix form, are

$$M^n = \begin{bmatrix} a^n & b^n - a^n \\ 0 & b^n \end{bmatrix},$$

thereby explaining the above phenomenon.

Quantum Mechanical Application

In quantum mechanical and quantum chemical computations matrix diagonalization is one of the most frequently applied numerical processes. The basic reason is that the time-independent Schrödinger equation is an eigenvalue equation, albeit in most of the physical situations on an infinite dimensional space (a Hilbert space). A very common approximation is to truncate Hilbert space to finite dimension, after which the Schrödinger equation can be formulated as an eigenvalue problem of a real symmetric, or complex Hermitian, matrix. Formally this approximation is founded on the variational principle, valid for Hamiltonians that are bounded from below. But also first-order perturbation theory for degenerate states leads to a matrix eigenvalue problem.

Diagonalization

Now suppose in $\mu_A(\theta)=(\theta-\lambda_1)^{p_1}\ldots(\theta-\lambda_k)^{p_k}$ we have $p_j = 1$ for $1 \le j \le k$ (which means then there are only linear factors in the minimal polynomial) then there are exactly k distinct eigenvalues, with n_j linearly independent eigenvectors for the eigenvalue λ_j, $n_1 + \ldots + n_k = n$, dim ker$(A - \lambda_j Id) = n_j$; these n_j's being exactly the ones obtained in $x_A(\theta)=(\theta-\lambda_1)^{n_1}\ldots(\theta-\lambda_k)^{n_k}$, 1 and $X=\oplus_{j=1}^{r}\ker(A-\lambda_j Id)$.

Since there are $n_1 + \ldots + n_k = n$ linearly independent eigen-vectors, let us arrange them in an ordered list $\{x_1,\ldots,x_n\}$ obtaining $Ax_i = \lambda_i x_i$, $1 \le i \le n$.

Then these n vectors being linearly independent form a basis x and forming the matrix

$P=\begin{bmatrix} x_1 & x_2 & \cdots & x_n\end{bmatrix}_{n\times n}$ ($\because$ cach x_i is a column vector)

we can write this as

$$a = PD\left(D = \text{diag}(\lambda_1,\ldots,\lambda_n)\right)$$

where a is the matrix for A to start with (in some basis b). But P is invertible (Since its columns are linearly independent) and thus $a = PDP^{-1}$ i.e. $D = P^{-1}aP$.

This means that by a change of basis, the matrix a has been represented in some new basis as a diag- onal matrix; put differently, there is some basis with respect to which we can have a diagonal matrix representing A, this diagonal matrix has the eigen values λ_i on its columns, each λ_i repeated as many times as it occurs in $x_A(\theta)$A. One says that A is diagonalizable, the matrix P being known as the modal matrix and D being known as the spectral matrix for the operator A or the matrix a.

Examples

Example 1: Take some polynomial $a(\theta)=\theta^n+a_{n-1}\theta^{n-1}+\ldots+a_1\theta+a_0$ of degree n(if we do have some $a_n \ne 1$, just divide by a_n which is possible since a_n must be non-zero with deg $(a(\theta)) = n$).

The matrix

$$\begin{bmatrix} 0 & 1 & 0 & \cdots & 0 \\ 0 & 0 & 1 & \cdots & 0 \\ \vdots & \vdots & \vdots & \vdots & \vdots \\ 0 & 0 & 0 & \cdots & 1 \\ -a_0 & -a_1 & -a_2 & \cdots & -a_{n-1} \end{bmatrix}$$

is called the companion matrix of a(θ). Then

$$\det(\lambda I - C_a) = \begin{vmatrix} \lambda & -1 & 0 & \cdots & 0 \\ 0 & \lambda & -1 & \cdots & 0 \\ \vdots & \vdots & \vdots & \vdots & \vdots \\ 0 & 0 & 0 & \cdots & -1 \\ a_0 & a_1 & a_2 & \cdots & \lambda + a_{n-1} \end{vmatrix}$$

$$= \begin{vmatrix} 0 & -1 & 0 & \cdots & 0 \\ 0 & \lambda & -1 & \cdots & 0 \\ \vdots & \vdots & \vdots & \vdots & \vdots \\ 0 & 0 & 0 & \cdots & -1 \\ a(\lambda) & a_1 & a_2 & \cdots & \lambda + a_{n-1} \end{vmatrix}$$

(Adding λ times the 2nd column, λ^2 times the 3rd column, λ^{n-1} times the last column to the first column)

$= (-1)^{n+1} a(\lambda)(-1)^{n-1} = a(\lambda)$ (expanded in terms of the first column)

So that x(θ) = a(θ). We also have μ(θ) = a(θ) (∵ the minor of a_o in ($\lambda I - C_a$) is $(-1)^{n-1}$ and thus d_{n-1}, the gcd of all minors of order (n – 1) in ($\lambda I - C_a$) is 1).

If λ_i is an eigenvalue, we have $x(\lambda_i) = 0$ hence $(\lambda_i)^n = -\left[a_{n-1}(\lambda_i)^{n-1} + \ldots + a_1\lambda_i + a_0\right]$ and thus,

for $x_i = \begin{bmatrix} 1 \\ \lambda_i \\ \vdots \\ (\lambda_i)^{n-1} \end{bmatrix}$ we find $C_a x_i = \lambda_i x_i == \begin{bmatrix} \lambda_i \\ \lambda_i^2 \\ \vdots \\ (\lambda_i)^{n-1} \end{bmatrix}$.

Verifying that x_i is an eigenvector or corresponding to λ_i. We then form the matrix $V = [x_1 \ldots x_n]_{n\times n}$ so that $C_a V = V\,\mathrm{diag}(\lambda_1, \ldots, \lambda_n)$.

Now $V = \begin{bmatrix} 1 & 1 & \cdots & 1 \\ \lambda_1 & \lambda_2 & \cdots & \lambda_n \\ \vdots & \vdots & \vdots & \vdots \\ \lambda_1^{n-1} & \lambda_2^{n-2} & \cdots & \lambda_n^{n-1} \end{bmatrix}$

which is Vandermonde matrix with determinant $\prod_{1 \le i < j \le n} (\lambda_j - \lambda_i)$. If all the λ_i are distinct, this determinant is not zero, the modal matrix V is invertible, and we have $V^{-1} C_a V = \text{diag}(\lambda_1, \ldots, \lambda_n)$, so that $C = C_a$ is diagonalizable.

Example 2: For a complex vector equipped with an inner product space, we ask additional question: given $X \xrightarrow{A} X$, $\dim X = n < \infty$, do we have an orthonormal set of n eigenvectors of A? If so, we call it a normal matrix. Then if $\{e_1, \ldots, e_n\}$ happens to be the orthonormal set with n eigenvectors, $\{e_1, \ldots, e_n\}$ is a basis and thus a normal matrix is certainly diagonalizable. Further, if $U = [e_1 \ \ldots \ e_n]$ (the n × n matrix with e_i as columns), we can write $a = UDU^{-1}$.

Bilinear Form

In mathematics, more specifically in abstract algebra and linear algebra, a bilinear form on a vector space V is a bilinear map $V \times V \to K$, where K is the field of scalars. In other words, a bilinear form is a function $B : V \times V \to K$ that is linear in each argument separately:

- $B(u + v, w) = B(u, w) + B(v, w)$ and $B(\lambda u, v) = \lambda B(u, v)$
- $B(u, v + w) = B(u, v) + B(u, w)$ and $B(u, \lambda v) = \lambda B(u, v)$

The definition of a bilinear form can be extended to include modules over a ring, with linear maps replaced by module homomorphisms.

When K is the field of complex numbers C, one is often more interested in sesquilinear forms, which are similar to bilinear forms but are conjugate linear in one argument.

Coordinate Representation

Let $V \cong K^n$ be an n-dimensional vector space with basis $\{e_1, \ldots, e_n\}$.

Define the $n \times n$ matrix A by $A_{ij} = B(e_i, e_j)$.

If the $n \times 1$ matrix x represents a vector v with respect to this basis, and analogously, y represents w, then:

$$B(v,w) = x^T Ay = \sum_{i,j=1}^{n} a_{ij} x_i y_j.$$

Suppose $\{f_1, ..., f_n\}$ is another basis for *V*, such that:

$[f_1, ..., f_n] = [e_1, ..., e_n]S$

where $S \in GL(n, K)$.

Now the new matrix representation for the bilinear form is given by: S^TAS.

Maps to the Dual Space

Every bilinear form *B* on *V* defines a pair of linear maps from *V* to its dual space V^*. Define $B_1, B_2: V \to V^*$ by

$$B_1(v)(w) = B(v, w)$$

$$B_2(v)(w) = B(w, v)$$

This is often denoted as

$$B_1(v) = B(v, \cdot)$$

$$B_2(v) = B(\cdot, v)$$

where the dot (·) indicates the slot into which the argument for the resulting linear functional is to be placed.

For a finite-dimensional vector space *V*, if either of B_1 or B_2 is an isomorphism, then both are, and the bilinear form *B* is said to be nondegenerate. More concretely, for a finite-dimensional vector space, non-degenerate means that every non-zero element pairs non-trivially with some other element:

$B(x, y) = 0$ for all $y \in V$ implies that $x = 0$ and

$B(x, y) = 0$ for all $x \in V$ implies that $y = 0$.

The corresponding notion for a module over a commutative ring is that a bilinear form is unimodular if $V \to V^*$ is an isomorphism. Given a finitely generated module over a commutative ring, the pairing may be injective (hence "nondegenerate" in the above sense) but not unimodular. For example, over the integers, the pairing $B(x, y) = 2xy$ is nondegenerate but not unimodular, as the induced map from $V = Z$ to $V^* = Z$ is multiplication by 2.

If *V* is finite-dimensional then one can identify *V* with its double dual V^{**}. One can then show that B_2 is the transpose of the linear map B_1 (if *V* is infinite-dimensional then B_2 is the transpose of B_1 restricted to the image of *V* in V^{**}). Given *B* one can define the *transpose* of *B* to be the bilinear form given by

$${}^tB(v, w) = B(w, v).$$

The left radical and right radical of the form B are the kernels of B_1 and B_2 respectively; they are the vectors orthogonal to the whole space on the left and on the right.

If V is finite-dimensional then the rank of B_1 is equal to the rank of B_2. If this number is equal to dim(V) then B_1 and B_2 are linear isomorphisms from V to V^*. In this case B is nondegenerate. By the rank–nullity theorem, this is equivalent to the condition that the left and equivalently right radicals be trivial. For finite-dimensional spaces, this is often taken as the *definition* of nondegeneracy:

Definition: B is nondegenerate if $B(v, w) = 0$ for all w implies $v = 0$.

Given any linear map $A : V \to V^*$ one can obtain a bilinear form B on V via

$$B(v, w) = A(v)(w).$$

This form will be nondegenerate if and only if A is an isomorphism.

If V is finite-dimensional then, relative to some basis for V, a bilinear form is degenerate if and only if the determinant of the associated matrix is zero. Likewise, a nondegenerate form is one for which the determinant of the associated matrix is non-zero (the matrix is non-singular). These statements are independent of the chosen basis. For a module over a commutative ring, a unimodular form is one for which the determinant of the associate matrix is a unit (for example 1), hence the term; note that a form whose matrix is non-zero but not a unit will be nondegenerate but not unimodular, for example $B(x, y) = 2xy$ over the integers.

Symmetric, Skew-symmetric and Alternating Forms

We define a bilinear form to be

- symmetric if $B(v, w) = B(w, v)$ for all v, w in V;
- alternating if $B(v, v) = 0$ for all v in V;
- skew-symmetric if $B(v, w) = -B(w, v)$ for all v, w in V;

  Proposition: Every alternating form is skew-symmetric.

  Proof: This can be seen by expanding $B(v + w, v + w)$.

If the characteristic of K is not 2 then the converse is also true: every skew-symmetric form is alternating. If, however, char(K) = 2 then a skew-symmetric form is the same as a symmetric form and there exist symmetric/skew-symmetric forms that are not alternating.

A bilinear form is symmetric (resp. skew-symmetric) if and only if its coordinate matrix (relative to any basis) is symmetric (resp. skew-symmetric). A bilinear form is alternating if and only if its coordinate matrix is skew-symmetric and the diagonal entries are all zero (which follows from skew-symmetry when char(K) $\neq$ 2).

A bilinear form is symmetric if and only if the maps B_1, B_2: $V \rightarrow V^*$ are equal, and skew-symmetric if and only if they are negatives of one another. If char(K) ≠ 2 then one can decompose a bilinear form into a symmetric and a skew-symmetric part as follows

$$B^+ = \tfrac{1}{2}(B + {}^t B) \qquad B^- = \tfrac{1}{2}(B - {}^t B),$$

where ${}^tB$ is the transpose of B (defined above).

Derived Quadratic Form

For any bilinear form $B : V \times V \rightarrow K$, there exists an associated quadratic form $Q : V \rightarrow K$ defined by $Q : V \rightarrow K : v \mapsto B(v, v)$.

When char(K) ≠ 2, the quadratic form Q is determined by the symmetric part of the bilinear form B and is independent of the antisymmetric part. In this case there is a one-to-one correspondence between the symmetric part of the bilinear form and the quadratic form, and it makes sense to speak of the symmetric bilinear form associated with a quadratic form.

When char(K) = 2 and dim $V > 1$, this correspondence between quadratic forms and symmetric bilinear forms breaks down.

Reflexivity and Orthogonality

Definition: A bilinear form $B : V \times V \rightarrow K$ is called reflexive if $B(v, w) = 0$ implies $B(w, v) = 0$ for all v, w in V.

Definition: Let $B : V \times V \rightarrow K$ be a reflexive bilinear form. v, w in V are orthogonal with respect to B if $B(v, w) = 0$.

A bilinear form B is reflexive if and only if it is either symmetric or alternating. In the absence of reflexivity we have to distinguish left and right orthogonality. In a reflexive space the left and right radicals agree and are termed the *kernel* or the *radical* of the bilinear form: the subspace of all vectors orthogonal with every other vector. A vector v, with matrix representation x, is in the radical of a bilinear form with matrix representation A, if and only if $Ax = 0 \Leftrightarrow x^T A = 0$. The radical is always a subspace of V. It is trivial if and only if the matrix A is nonsingular, and thus if and only if the bilinear form is nondegenerate.

Suppose W is a subspace. Define the *orthogonal complement*

$$W^\perp = \{v \mid B(v, w) = 0 \ \forall w \in W\}.$$

For a non-degenerate form on a finite dimensional space, the map $W \mapsto W^\perp$ is bijective, and the dimension of $W^\perp$ is $\dim(V) - \dim(W)$.

Different Spaces

Much of the theory is available for a bilinear mapping from two vector spaces over the same base field to that field

$$B : V \times W \to K.$$

Here we still have induced linear mappings from V to W^*, and from W to V^*. It may happen that these mappings are isomorphisms; assuming finite dimensions, if one is an isomorphism, the other must be. When this occurs, B is said to be a perfect pairing.

In finite dimensions, this is equivalent to the pairing being nondegenerate (the spaces necessarily having the same dimensions). For modules (instead of vector spaces), just as how a nondegenerate form is weaker than a unimodular form, a nondegenerate pairing is a weaker notion than a perfect pairing. A pairing can be nondegenerate without being a perfect pairing, for instance $Z \times Z \to Z$ via $(x,y) \mapsto 2xy$ is nondegenerate, but induces multiplication by 2 on the map $Z \to Z^*$.

Terminology varies in coverage of bilinear forms. For example, F. Reese Harvey discusses "eight types of inner product". To define them he uses diagonal matrices A_{ij} having only +1 or −1 for non-zero elements. Some of the "inner products" are symplectic forms and some are sesquilinear forms or Hermitian forms. Rather than a general field K, the instances with real numbers R, complex numbers C, and quaternions H are spelled out. The bilinear form

$$\sum_{k=1}^{p} x_k y_k - \sum_{k=p+1}^{n} x_k y_k$$

is called the real symmetric case and labeled $R(p, q)$, where $p + q = n$. Then he articulates the connection to traditional terminology:

> Some of the real symmetric cases are very important. The positive definite case $R(n, 0)$ is called *Euclidean space*, while the case of a single minus, $R(n-1, 1)$ is called *Lorentzian space*. If $n = 4$, then Lorentzian space is also called *Minkowski space* or *Minkowski spacetime*. The special case $R(p, p)$ will be referred to as the *split-case*.

Relation to Tensor Products

By the universal property of the tensor product, bilinear forms on V are in 1-to-1 correspondence with linear maps $V \otimes V \to K$. If B is a bilinear form on V the corresponding linear map is given by

$$v \otimes w \mapsto B(v, w)$$

The set of all linear maps $V \otimes V \to K$ is the dual space of $V \otimes V$, so bilinear forms may be thought of as elements of

$$(V \otimes V)^* \cong V^* \otimes V^*$$

Likewise, symmetric bilinear forms may be thought of as elements of $\mathrm{Sym}^2(V^*)$ (the second symmetric power of V^*), and alternating bilinear forms as elements of $\Lambda^2 V^*$ (the second exterior power of V^*).

On normed Vector Spaces

Definition: A bilinear form on a normed vector space $(V, \|\cdot\|)$ is bounded, if there is a constant C such that for all u, $\mathbf{v} \in V$,

$$B(u,v) \leq C\|u\|\|v\|.$$

Definition: A bilinear form on a normed vector space $(V, \|\cdot\|)$ is elliptic, or coercive, if there is a constant $c > 0$ such that for all $u \in V$,

$$B(u,u) \geq c\|u\|^2.$$

Generalization to Modules

Given a ring R and a right R-module M and its dual module M^*, a mapping $B : M^* \times M \to R$ is called a bilinear form if

$$B(u + v, x) = B(u, x) + B(v, x)$$

$$B(u, x + y) = B(u, x) + B(u, y)$$

$$B(\alpha u, x\beta) = \alpha B(u, x)\beta$$

for all $u, v \in M^*, x, y \in M, \alpha, \beta \in R$.

The mapping $\langle\cdot,\cdot\rangle : M^* \times M \to R : (u, x) \mapsto u(x)$ is known as the *natural pairing*, also called the *canonical bilinear form* on $M^* \times M$.

A linear map $S : M^* \to M^* : u \mapsto S(u)$ induces the bilinear form $B : M^* \times M \to R : (u, x) \mapsto \langle S(u), x\rangle$, and a linear map $T : M \to M : x \mapsto T(x)$ induces the bilinear form $B : M^* \times M \to R : (u, x) \mapsto \langle u, T(x)\rangle\rangle$. Conversely, a bilinear form $B : M^* \times M \to R$ induces the R-linear maps $S : M^* \to M^* : u \mapsto (x \mapsto B(u, x))$ and $T' : M \to M^{**} : x \mapsto (u \mapsto B(u, x))$. Here, M^{**} denotes the double dual of M.

More on Sesqui/Bi-Linear Forms

1. Consider now a sesquiliner form $X \times X \xrightarrow{s} \mathbb{K}$ and introduce $s^*(x,u) := \overline{s(u,x)}$.

Then s^* again a sesquilinear form$\left(s^*(x\lambda,u\mu):=\overline{s(u\mu,x\lambda)}=\overline{\mu}\overline{s(u,x)}\lambda=\overline{\lambda}\overline{s(u,x)}\mu=\overline{\lambda}s^*(x,u)\mu\right)$.

Let $Y\in \mathrm{Cen}\mathbb{K}$; this means $Y\lambda = \lambda Y$ for each $\lambda \in \mathbb{K}$ and hence $\overline{Y\lambda}=\overline{\lambda}\overline{\gamma}=\overline{\gamma}\overline{\lambda}=\lambda\overline{\gamma}$ for each $\lambda\in\mathbb{K}$. Thus $Y \in \mathrm{Cen}(\mathbb{K})$ iff $\overline{\gamma}\in \mathrm{Cen}\mathbb{K}$. Let $\Gamma(\mathbb{K}):= y \in \mathrm{Cen}(\mathbb{K})\,|\,y\overline{y}=1$. Then $\Gamma(\mathbb{K})$ is not empty since $\pm1\in\Gamma(\mathbb{K})$. We make a definition. A sesquilinear $X\times X\xrightarrow{S}\mathbb{K}$ is called y–hermitian ($y\in\Gamma(\mathbb{K})$ iff $s=\gamma s^*$, i.e, $s(x,u)=\gamma\overline{s(x,u)}$ at each $x, u \in X$.

(i) When $\gamma = 1$, we say that γ-hermitian form is hermitian, when $\gamma = -1$, we say it is skew-hermitian. Thus s is hermitian iff $s(x,u)=\overline{s(x,u)}$ and s is skew hermitian iff $s(x,u)=-\overline{s(x,u)}$.

(ii) For the trivial involution (thus 1 is forced to be commutative) we have bilinear forms and 'hermitian' is called 'symmetric'. Thus a bilinear form β is called symmetric iff β (x, u) = β (u, x) and it is called skew symmetric iff β (x, u) = $-\beta$ (u, x).

2. When $X\xrightarrow{\tau}X$ is an isomorphism with $s(\tau x,\tau)=s(x,u)$, it is called an isometry on X; the term unitary transformation is also used. The unitary transformation is also used. The unitary transformations of X constitute a group under composition $\left(s\left(\tau^{-1}x,\tau u\right)=s\left(\tau\left(\tau^{-1}x\right),\tau\left(\tau^{-1}u\right)\right)=s(x,u)\right)$ with id_X as the identity. (A group is a set A with an associative binary operation (i.e. a (bc) = (ab) c holds) in which each a $\in$ A has an inverse a^{-1} so that $aa^{-1} = 1 = a^{-1}a$; the existence of an identity is part of the definition. The nonzero quaternions from a group; in fact the nonzero elements of any division ring do. The eight quaternions $\pm e_i \mid i = 0, 1, 2, 3$ also form a group; it is called the quaternion group. When a group is commutative, it is called 'abelian'; the additive fragment of any ring is an abelian group, the nonzero elements of a field form an abelian group).

3. (i) Let $X\times X\xrightarrow{S}\mathbb{K}$ be a sesquilinear form. Then if $\mathbb{C}$ is a subring of $\mathbb{K}$ and the involution on $\mathbb{K}$ becomes involution of complex numbers i.e. conjugation on $\mathbb{C}$, we have (since $\overline{i}=i^{-1},\overline{i^k}=i^{-k}$ and s is conjugate-linear in the first variable).

$$\sum_{k=0}^{3} s(x+ui^k,x+ui^k)i^k = \sum_{k=0}^{3}[s(x,x)+$$

$$s(x,u)i^k+s(u,x)i^{-k}+s(u,u)i^{-k}i^k]i^k = \sum_{k=0}^{3}[s(x,x)i^k+s(x,u)i^{2k}+s(u,x)i^{-k}i^k+s(u,u)i^k] =$$

$$\sum_{k=0}^{3}[s(x,x)-s(x,u)+s(u,u)]i^k+\sum_{k=0}^{3}s(u,x) = [s(x,x)-s(x,u)+s(u,u)][1+i+i^2+i^3]+$$

$4s(u,x) = 4s(u,x)$ ($\because 1+i+i^2+i^3=0$) The relation $4s(u,x) = \sum_{k=0}^{3} s(x+ui^k,x+ui^k)i^k$

is known as the Polarization identity for the sesquilinear forms. (ii) The rings $\mathbb{K}=\mathbb{C},\mathbb{K}=\mathbb{H}$ (the quaternions) clearly satisfy the requirement for $\mathbb{K}$ in (i) above. There are in fact many other $\mathbb{K}$ (e.g the ring of complex n × m matrices with conjugation on $\mathbb{K}$ defined as the conjugate-transposse) for which the calculations in (i) are valid but we will not discuss the more general situation (which needs $\mathbb{K}$ to be what is known as a C*-algebra).

4. (i) If we write s(x, x) as Q(x), we associate a function $X \xrightarrow{X} \mathbb{K}$ called the quadratic form of s. Then the polarization identity says that $s(u,x)=\frac{1}{4}\left[\sum_{k=0}^{3}Q\left(x+ui^k,x+ui^k\right)i^k\right]$ so that the sesquilinear form is completely determined by its quadratic form.

(ii) Noting that $s^*(u,x):=\overline{s(u,x)}$ is also sesquilinear and the quadratic form for $s^*(u,x)$ is then $Q^*(x):=\overline{Q(x)}$ we conclude that s is γ-hermitian iff $Q^*(x):=\overline{Q(x)}=\gamma Q(x)$. In particular: A sesquilinear form s is hermitian/skew hermitian iff its associated quadratic form satisfies $Q(x):=\overline{Q(x)}/Q(x)=-\overline{Q(x)}$.

(iii) When $\mathbb{K}=\mathbb{C}\text{ or }\mathbb{H}$, we conclude by using (ii) ($\mathbb{H}$ =quaternions) A sesquilinear form $X\times X \xrightarrow{s} \mathbb{K}(\mathbb{K}=\mathbb{C}\text{ or }\mathbb{H})$ is hermitian iff Q(x) ∈ $\mathbb{R}$.

(iv) Suppose $\mathbb{K}=\mathbb{C}\text{ or }\mathbb{H}$. Let the real part of a ∈ $\mathbb{K}$ be denoted by a_o (thus

$$a=\begin{cases} a_0+a_1 i \text{ for } a\in\mathbb{K} \\ \quad a_0+ia_1 \end{cases}$$

and a = a_o + a where a = $e_1a_1 + e_2a_2 + e_3a_3$ for a ∈ $\mathbb{H}$) Then a + a = $2a_o$. If s is hermitian, we have

$$s(x\lambda+u\mu,x\lambda+u\mu)=s(x\lambda,x\lambda)+s(u\mu,x\lambda)+s(x\lambda,u\mu)+s(u\mu,umu)$$

$$=\overline{\lambda}s(x,x)\lambda+\overline{\mu}s(u,x)\lambda+\overline{\lambda}s(x,u)\mu+\overline{\mu}s(u,u)\mu$$

$$=\overline{\lambda}s(x,x)\lambda+\overline{\overline{\lambda}s(x,u)\mu}+\overline{\lambda}s(x,u)\mu+\overline{\mu}s(u,u)\mu$$

$$(\because \overline{\overline{\lambda}s(x,u)\mu}=\overline{\mu}\,\overline{s(x,u)}\,\overline{\overline{\lambda}}=\overline{\mu}s(u,x)\lambda)$$

$=|\lambda|^2s(x,x)+2(\overline{\lambda}s(x,u)\mu)+|\mu|^2s(u,u)$ ($\because s(z,z)\in\mathbb{R}$ for each $z\in X$ and $\mathbb{R}=Cen(\mathbb{H})$;

for $\mathbb{K}=\mathbb{C}$ the argument is obvious any way)

That is, $Q(x\lambda+u\mu)=|\lambda|^2Q(x)+2(\overline{\lambda}s(x,u)\mu)_0+|\mu|^2Q(u)$ *(every term in this is a real number)*

(v) For complex numbers $a = a_0 + ia_1$ will be written $a = \underline{re}(a) + i\overline{im}(a)$. Then $ai = a0i - a_1$ so $\underline{re}(ai) = -im(a)$ and $\underline{im}(ai) = re(a)$. For $X \times X \xrightarrow{s} \mathbb{C}$,

we have $s(xi, u) = \bar{i}s(x, u) = -is(x, u)$. Thus $\underline{ims}(x, u) = -\underline{re}(is(x, u)) = \underline{res}(xi, u)$.

Further, s is hermitian iff $s(x, u) := \overline{s(u, x)}$ and hence we have, from $s(xi, u) = \bar{i}s(x, u)$, s is hermitian iff $s(xi, u) = \bar{i}\overline{s(u, x)} = \overline{s(u, x)i} = -i\overline{s(u, x)} = -is(x, u)$ i.e. s is hermitian iff is is skew hermitian.

References

- Horn, Roger A.; Johnson, Charles R. (2013). Matrix Analysis (Second Edition). Cambridge University Press. ISBN 978-0-521-83940-2
- Halmos, Paul R. (1974). Finite-dimensional vector spaces. Undergraduate Texts in Mathematics. Berlin, New York: Springer-Verlag. ISBN 978-0-387-90093-3. Zbl 0288.15002
- Milnor, J.; Husemoller, D. (1973). Symmetric Bilinear Forms. Ergebnisse der Mathematik und ihrer Grenzgebiete. 73. Springer-Verlag. ISBN 3-540-06009-X. Zbl 0292.10016
- Porteous, Ian R. (1995). Clifford Algebras and the Classical Groups. Cambridge Studies in Advanced Mathematics. 50. Cambridge University Press. ISBN 978-0-521-55177-9
- Cooperstein, Bruce (2010). "Ch 8: Bilinear Forms and Maps". Advanced Linear Algebra. CRC Press. pp. 249–88. ISBN 978-1-4398-2966-0

Permissions

All chapters in this book are published with permission under the Creative Commons Attribution Share Alike License or equivalent. Every chapter published in this book has been scrutinized by our experts. Their significance has been extensively debated. The topics covered herein carry significant information for a comprehensive understanding. They may even be implemented as practical applications or may be referred to as a beginning point for further studies.

We would like to thank the editorial team for lending their expertise to make the book truly unique. They have played a crucial role in the development of this book. Without their invaluable contributions this book wouldn't have been possible. They have made vital efforts to compile up to date information on the varied aspects of this subject to make this book a valuable addition to the collection of many professionals and students.

This book was conceptualized with the vision of imparting up-to-date and integrated information in this field. To ensure the same, a matchless editorial board was set up. Every individual on the board went through rigorous rounds of assessment to prove their worth. After which they invested a large part of their time researching and compiling the most relevant data for our readers.

The editorial board has been involved in producing this book since its inception. They have spent rigorous hours researching and exploring the diverse topics which have resulted in the successful publishing of this book. They have passed on their knowledge of decades through this book. To expedite this challenging task, the publisher supported the team at every step. A small team of assistant editors was also appointed to further simplify the editing procedure and attain best results for the readers.

Apart from the editorial board, the designing team has also invested a significant amount of their time in understanding the subject and creating the most relevant covers. They scrutinized every image to scout for the most suitable representation of the subject and create an appropriate cover for the book.

The publishing team has been an ardent support to the editorial, designing and production team. Their endless efforts to recruit the best for this project, has resulted in the accomplishment of this book. They are a veteran in the field of academics and their pool of knowledge is as vast as their experience in printing. Their expertise and guidance has proved useful at every step. Their uncompromising quality standards have made this book an exceptional effort. Their encouragement from time to time has been an inspiration for everyone.

The publisher and the editorial board hope that this book will prove to be a valuable piece of knowledge for students, practitioners and scholars across the globe.

Index